Lecture Notes in Mathematics

Edited by A. Dold and B. Eckmann

466

Non-Commutative Harmonic Analysis

Actes du Colloque d' Analyse Harmonique Non Commutative, Marseille-Luminy, 1 au 5 Juillet 1974

Edited by J. Carmona, J. Dixmier and M. Vergne

Springer-Verlag
Berlin · Heidelberg · New York 1975

Editors
Prof. Jacques Carmona
Université d' Aix-Marseille
Départment de Mathématiques
70 Route Léon Lachamp
13288 Marseille Cedex 2/France

Prof. Jaques Dixmier
Université Paris VI
U.E.R. d' Analyse
Probalité et applications
4 Place Jussieu
75230 Paris Cedex 05/France

Prof. Michèle Vergne
Université Paris VII
U.E.R. de Mathématiques
2 Place Jussieu
75221 Paris Cedex 05/France

Library of Congress Cataloging in Publication Data

Colloque d'analyse harmonique non commutative, Marseille, 1974.
Non commutative harmonic analysis.

(Lecture notes in mathematics (Berlin) ; 466)
English or French.
1. Harmonic analysis--Congresses. 2. Lic algebras--Congresses. 3. Locally compact groups--Congresses. I. Carmona, Jacques, 1934-
II. Dixmier, Jacques. III. Vergne, Michèle.
IV. Title. V. Series.
QA3.L28 no.466 [QA403] 515'.785 75-19252

AMS Subject Classifications (1970): 16A66, 17B20, 17B35, 17B45, 20G05, 22D10, 22D12, 22E45, 22E50, 31A10, 35P15, 43A05, 43A65, 82A15

ISBN 3-540-07183-0 Springer-Verlag Berlin · Heidelberg · New York
ISBN 0-387-07183-0 Springer-Verlag New York · Heidelberg · Berlin

Printed in Germany

Offsetdruck: Julius Beltz, Hemsbach/Bergstr.

PREFACE

Un colloque d'Analyse Harmonique Non Commutative a eu lieu à Marseille-Luminy, du 1 au 5 juillet 1974, dans le cadre des activités du Centre International de Rencontres Mathématiques avec le soutien de l'U. E. R. de Luminy (Université d'Aix-Marseille).

Le présent volume contient le texte des exposés que les conférenciers invités ont bien voulu nous faire parvenir. La liste des articles ne coïncide pas exactement avec celle des exposés présentés durant le Colloque. C'est le cas en particulier des conférences de Kostant, Raïs, Wallach.

Outre les participants à cette rencontre, nous tenons à remercier l'U. E. R. de Luminy et le Centre International de Rencontres Mathématiques qui ont rendu possible la tenue de ce colloque, ainsi que le secrétariat du Département de Mathématique-Informatique de Luminy qui a assuré la préparation matérielle de ce volume.

Jacques CARMONA
Jacques DIXMIER
Michèle VERGNE

TABLE DES MATIERES

INTERTWINING OPERATORS AND THE HALF-DENSITY PAIRING

Robert J. Blattner

1. Introduction.

These notes give the details of a result stated in a previous paper ([2], Section 7(b)) which asserts that the Knapp-Stein intertwining operator [3] linking the principal series representations of SL(2,ℝ) coming from characters of a split Borel subgroup differing by a Weyl reflection can be constructed from the so-called half-density pairing ([2], p. 152), due to Kostant and Sternberg, of geometric quantization.

Section 2 of the present paper recalls a construction of [4], which manufactures a hermitian line bundle with connection (L,∇) given an orbit X of the coadjoint representation of a simply connected Lie group G and a point $p \in X$. We also derive some results relating non-vanishing sections of L and certain 1-forms they define. The principal result here is Proposition 2.6. Section 3 is devoted to working out our example. G is the universal covering group of SL(2,ℝ) and X is a hyperboloid of one sheet. Whereas Knapp and Stein fix a Borel subgroup of G and let the Weyl group act on the inducing representation, the geometry of our situation leads us to use the Weyl group to move the Borel subgroup while holding the inducing representation fixed. Section 4 comments on our result and raises questions that need to be settled concerning the half-density pairing.

In what follows we shall assume the reader to be familiar with Kostant's fundamental paper [4] on geometric pre-quantization and also with the half-density pairing as set forth in [2].

This work was supported in part by NSF Grant GP-43376.

2. Remarks on the orbit method.

Let G be a connected, simply connected Lie group with Lie algebra g. Let g^* be the dual of g and let ad* denoted the coadjoint representation of G on g^*: $\mathrm{ad}^*\, x = ({}^t \mathrm{ad}\, x)^{-1}$. One constructs representations of G by the orbit method by choosing an orbit X of ad* in g^*, choosing an ad*-invariant polarization F of X, and constructing a Hilbert space upon which G acts from these data. For details, see Auslander and Kostant ([1], Section I.5).

Suppose F is real. Then the construction in ([2], Section 3) is equivalent to that given in [1]. We use this alternative construction in the present paper. Let $p \in X$ and let $G_p = \{x \in G : (\mathrm{ad}^* x)p = p\}$. As detailed in ([4], pp. 197-199), we may construct a complex hermitian line bundle with connection (L,∇) as follows: Let χ be any unitary character of G_p such that $d\chi = 2\pi i p$ (if one exists). Form the complex hermitian line bundle $G \times \mathbb{C}$ over G using the usual hermitian structure on $\mathbb{C}$. Let G_p act on $G \times \mathbb{C}$ on the right by means of

$$(2.1) \qquad (x,\lambda)y = (xy,\chi(y)^{-1}\lambda)$$

and on G by right translation. Then $G \times \mathbb{C}/G_p$ becomes a complex hermitian line bundle L over G/G_p. G/G_p is identified with X in the usual way by means of the map $\pi : x \mapsto (\mathrm{ad}^* x)p$ of G onto X. Let π also denote the canonical projection of $G \times \mathbb{C}$ onto L.

Let α_p be the left invariant 1-form on G whose value at 1 is p. Let $\mathbb{C}^\times = \mathbb{C} - \{0\}$ and let $L^\times = L - \{0\text{-section}\}$. Then ([4], p. 199) the 1-form $(\alpha_p, \frac{1}{2\pi i}\frac{dz}{z})$ on $G \times \mathbb{C}^\times$ is $\pi^*\alpha$ for a unique connection form α on $L^\times$, and the connection ∇ associated to α by means of

$$(2.2) \qquad \nabla_\xi s = 2\pi i \langle s^*\alpha, \xi\rangle s(q),$$

where $q \in X$, s is a section of $L^\times$ near q, and $\xi \in (TX)_q$, leaves the hermitian structure of L invariant.

Let f be a $\mathbb{C}$-valued function on G. Then there is a section s_f of L over X such that $\pi(x,f(x)) = s_f(\pi(x))$ if and only if

(2.3) $$f(x,y) = \chi(y)^{-1}f(x) \text{ for } x \in G,\ y \in G_p,$$

and in that case s_f determines f uniquely. Moreover s_f never vanishes if and only if f never vanishes, and in that case $\pi^*s_f^*\alpha = \phi^*(\alpha_p, \frac{1}{2\pi i}\frac{dz}{z})$, where $\phi(x) = (x,f(x))$. Setting $\alpha(f) = s_f^*\alpha$, we get

(2.4) $$\pi^*\alpha(f) = \alpha_p + \frac{1}{2\pi i}\frac{df}{f}.$$

(Note that neither α_p nor $\frac{1}{2\pi i}\frac{df}{f}$ come from 1-forms on X via $\pi : G \to X$ although their sum does.)

Which 1-forms β on X are of the form $\alpha(f)$ for a smooth never-vanishing function f on G? Clearly we must have $d\beta = \omega$, where ω is the canonical symplectic 2-form on X due to Kirillov ([4], p. 182). Moreover we must have, by (2.3) and (2.4),

(2.5) $$\chi(y) = e^{-2\pi i \int_{\hat{\gamma}}(\pi^*\beta - \alpha_p)} \text{ for } y \in G_p.$$

where $\hat{\gamma}$ is any piecewise smooth arc in G from 1 to y. The following proposition says that this is sufficient.

<u>Proposition</u> 2.6: Let β be a smooth 1-form on X such that $d\beta = \omega$. Then $\pi^*\beta - \alpha_p$ is closed on G, so that integrations of it are independent of path. Set

(2.7) $$f(x) = e^{2\pi i \int_1^x (\pi^*\beta - \alpha_p)} \text{ for } x \in G_p,$$

where the integration is over any arc from 1 to x. Then

(a) $(f|G_p)^{-1}$ is a character Λ of G_p,

(b) $d\Lambda = 2\pi i p$, and

(c) $f(xy) = \Lambda(y)^{-1}f(x)$ for $x \in G$, $y \in G_p$.

<u>Proof</u>: To show that $\pi^*\beta - \alpha_p$ is closed, we show that $d\alpha_p = \pi^*\dot{\omega}$. Let ξ, η be two left invariant vector fields on G. Then according to ([4], p. 96) $d\alpha_p(\xi,\eta) = \langle\alpha_p,[\eta,\xi]\rangle$, since $\langle\alpha_p,\xi\rangle$ and $\langle\alpha_p,\eta\rangle$ are constant functions. But this is just the definition of $\pi^*\omega$. Since G is simply connected, integrations of $\pi^*\beta - \alpha_p$ are independent of path.

Our next step is to show

$$(2.8)\quad \Lambda(y) = e^{-2\pi i\int_x^{xy}(\pi^*\beta-\alpha_p)} \quad \text{for all } x \in G,\ y \in G_p.$$

This will imply that

$$f(xy) = e^{2\pi i\int_1^{x}(\pi^*\beta-\alpha_p)}\, e^{2\pi i\int_x^{xy}(\pi^*\beta-\alpha_p)} = \Lambda(y)^{-1}f(x)$$

for $x \in G$, $y \in G_p$, which gives (c) and, specializing x to G_p, (a) as well. Now $\int_x^{xy}\alpha_p$ is obviously independent of x since α_p is left invariant. So let $\tilde{\gamma}$ be any arc in G from 1 to $y \in G_p$ and set $\gamma = \pi\circ\tilde{\gamma}$. γ is closed. Set $\phi(x) = \int_{(ad^*x)^{-1}\gamma}\beta$ for $x \in G$ and let $\xi \in \mathfrak{g}$. ξ defines a vector field, also called ξ, on X by means of

$$(\xi\psi)(q) = \frac{d}{dt}\psi((ad^*\exp t\xi)^{-1}q)\big|_{t=0}$$

for $\psi \in C(X)$ and $q \in X$. Then

$$(2.9)\qquad (\xi\phi)(x) = \int_{(ad^*x)^{-1}\gamma}\theta(\xi)\beta \quad \text{for } x \in G,$$

where $\theta(\xi)$ denotes the Lie derivative with respect to ξ. Now $\theta(\xi)\beta = di(\xi)\beta + i(\xi)d\beta$, where $i(\xi)$ is the left interior product with respect to ξ. The first term is exact. The second term is just $i(\xi)\omega$, which is also exact by ([4], Proposition 5.3.1). Since the integral in (2.9) is over a closed arc, it follows that $\xi\phi = 0$ for all $\xi \in \mathfrak{g}$. Since G is connected, ϕ is constant, which proves (2.8).

As to (b), let $y \in (G_p)_0$, the identity component of G_p, and for this y let $\tilde{\gamma}$ be as above, with $\tilde{\gamma}$ in $(G_p)_0$. Then $\gamma = \pi\circ\tilde{\gamma}$ is a constant arc, which implies that

(2.10) $\Lambda(y) = e^{2\pi i \int_{\tilde{\gamma}} \alpha_p}$ for $y \in (G_p)_0$.

Differenting (2.10) with respect to $\xi \in g_p$, the Lie algebra of G_p, and evaluating at $y = 1$ gives (b).

We close this section by recalling two facts relating sections s_f and their corresponding 1-forms $\alpha(f)$. Firstly, let f_1 and f_2 be two never-vanishing functions satisfying (2.3). Then ([4], Proposition 1.9.1) implies

(2.11) $\langle s_{f_2}, s_{f_1} \rangle \, e^{2\pi i \phi}$, where $d\phi = \alpha(f_2) - \overline{\alpha(f_1)}$.

In particular, $\|s_f\|$ is constant if and only if $\alpha(f)$ is real. Secondly, let F be a polarization of X. Then s_f is covariant constant with respect to F if and only if $\langle \alpha(f), F \rangle = 0$.

3. The example.

Let G be the simply connected covering group of $SL(2,\mathbb{R})$. g consists of all matrices of the form $\begin{bmatrix} a & b \\ c & -a \end{bmatrix}$ with $a,b,c \in \mathbb{R}$, which we will denote by (a,b,c). We have the commutation rule

(3.1) $[(a,b,c),(a',b',c')] = (bc'-cb',\ 2(ab'-ba'), 2(ca'-ac'))$.

g possesses an ad G invariant symmetric bilinear form B defined by

(3.2) $B((a,b,c),(a',b',c')) = aa' + \frac{1}{2}(bc' + cb')$.

Using B to identify g with g^*, we replace ad* by ad in the orbit method.

Let us look at the orbit $X = \{\xi \in g : B(\xi,\xi) = \lambda^2\}$, where $\lambda \neq 0$. We calculate the Kirillov form ω on X.

Lemma 3.3: $\omega = (2b)^{-1}da\wedge db = (4a)^{-1}db\wedge dc = (2c)^{-1}dc\wedge da$ on those portions of X where these expressions make sense.

Proof: Transferring the definition of ω ([4], p. 182) to $\mathfrak{g}$, we obtain $\omega_p([\xi,p],[\eta,p]) = B(p,[\eta,\xi])$, where $p \in X$ and $\xi,\eta \in \mathfrak{g}$. But this equals $B([\xi,p]),\eta)$. Let $p = (a,b,c)$ with $b \neq 0$. We have

$$[(0,0,-b^{-1}),(a,b,c)] = (1,0,-2ab^{-1}) = \xi', \text{ say, and}$$

$$[((2b)^{-1},0,0),(a,b,c)] = (0,1,-cb^{-1}) = \eta'.$$

Then $\omega_p(\xi',\eta') = B((1,0,-2ab^{-1}),((2b)^{-1},0,0)) = (2b)^{-1}$. Therefore $\omega = (2b)^{-1}da\wedge db$ on that part of X where $b \neq 0$. The other formulae follow since $2ada + bdc + cdb = 0$ on X.

Let us apply the methods of Section 2 to this setting. Set $p = (\lambda,0,0)$. Then G_p is the direct product of $\exp\{(a,0,0):a \in \mathbb{R}\}$ and the center Z of G. Z is infinite cyclic with generator z, where z is the image of π in the lift $k(\cdot)$ of the one-parameter subgroup $t \mapsto \begin{bmatrix} \cos t & \sin t \\ -\sin t & \cos t \end{bmatrix}$ of $SL(2,\mathbb{R})$ to G. Thus the characters of G_p which give the homogeneous line bundles L over X of Section 2 are given as follows: Let $|s| = 1$. Then define the character χ_s of G_p by

$$\chi_s(\exp(a,0,0)) = e^{2\pi i\lambda a} \quad \text{and} \tag{3.4}$$

$$\chi_s(z) = s. \tag{3.5}$$

Now there are exactly two real polarizations F_1 and F_2 of X which are ad G invariant. They are given by the generators of the one-sheeted hyperboloid X. Moreover, F_1 and F_2 both satisfy the Pukanszky condition ([5], pp. 15-16). It will be convenient to introduce coordinates (u,v) on X such that the leaves of F_1 are the curves u = constant and the leaves of F_2 are the curves v = constant. We set

(3.6) $u = \frac{a+\lambda}{b} = \frac{-c}{a-\lambda}$ and

(3.7) $v = \frac{a-\lambda}{b} = \frac{-c}{a+\lambda}$.

The meaning of these particular coordinates will become clearer below, expecially at (3.19). These coordinates are valid on $X' = \{(a,b,c) \in X : b \neq 0\}$. We have the inversion formulae

(3.8) $a = \lambda\frac{u+v}{u-v}, \quad b = \frac{2\lambda}{u-v}, \quad c = \frac{-2\lambda uv}{u-v}$,

which show that X' maps bijectively onto $\{(u,v) \in \mathbb{R}^2 : u \neq v\}$. In these coordinates

(3.9) $\omega = \lambda(u-v)^{-2}du\wedge dv.$

Since $H^2(X,\mathbb{Z}) = 0$, the line bundle L^s determined by χ_s is trivial. Thus L^s has global never-vanishing sections, which we may construct by looking for 1-forms β satisfying $d\beta = \omega$ and also (2.5) with $\chi = \chi_s$ and $y = z$, thanks to Proposition 2.6. Now in our case, (2.5) simplifies because the one-parameter subgroup $k(\cdot)$ defined in the paragraph before last has the property that $B(p,k'(0)) = 0$, so that $k^*\alpha_p = 0$. Thus (2.5) becomes

(3.10) $\int_\gamma \beta = -\frac{1}{2\pi i}\log s \pmod{\mathbb{Z}}$,

where $\gamma(t) = \text{ad}\, k(t)\cdot p = (\lambda \cos 2t, -\lambda \sin 2t, -\lambda \sin 2t)$ for $0 \leq t \leq \pi$.

Lemma 3.11: Choose $r \in \mathbb{R}$ so that $e^{2\pi ir} = s$. Then

$$\beta_1' = -\lambda[(u-v)^{-1}du-u(1+u^2)^{-1}du]-r\pi^{-1}(1+u^2)^{-1}du \quad \text{and}$$

$$\beta_2' = -\lambda[(u-v)^{-1}dv+v(1+v^2)^{-1}dv]-r\pi^{-1}(1+v^2)^{-1}dv$$

are the restrictions to X' of 1-forms β_1 and β_2, respectively, defined and real analytic on all of X. Moreover $d\beta_i = \omega$, $\langle\beta_i,F_i\rangle = 0$, and β_i satisfies (3.10) for $i = 1,2$.

Proof: Let $X^{\pm} = \{(a,b,c) \in X : \pm\lambda a > 0\}$. Then $X = X' \cup X^+ \cup X^-$. Next note that b and c are global coordinates on X^+ and on X^-. Now we show that $(1+u^2)^{-1}$ and $u^2(1+u^2)^{-1}$ extend to real analytic functions defined on all of X. Indeed, on $X' \cap X^+$, $(1+u^2)^{-1} = b^2(b^2+[a+\lambda]^2)^{-1}$ and $u^2(1+u^2)^{-1} = (a+\lambda)^2(b^2+[a+\lambda]^2)^{-1}$, whereas on $X' \cap X^-$, $(1+u^2)^{-1} = (a-\lambda)^2(c^2+[a-\lambda]^2)^{-1}$ and $u^2(1+u^2)^{-1} = c^2(c^2+[a-\lambda]^2)^{-1}$. These formulae extend $(1+u^2)^{-1}$ and $u^2(1+u^2)^{-1}$ to all of X^+ and X^-, and thus to all of X, in a real analytic way. Similarly, $(1+v^2)^{-1}$ and $v^2(1+v^2)^{-1}$ extend to X.

An easy computation shows that $db = 2\lambda(u-v)^{-2}(dv - du)$ and $dc = 2\lambda(u-v)^{-2}(-u^2dv + v^2du)$ on X'. Therefore, using (3.8),

$$dc + u^2db = -2\lambda(u^2-v^2)(u-v)^{-2}du = -2adu. \tag{3.12}$$

We also note that on X'

$$-\lambda\left[\frac{1}{u-v} - \frac{u^2}{1+u^2}\right] = \frac{-\lambda(1+uv)}{(u-v)(1+u^2)} = \frac{c-b}{2(1+u^2)} \tag{3.13}$$

Combining (3.12) and (3.13) we get that β_1' is given on $X' \cap (X^+ \cup X^-)$ by

$$-\frac{1}{2a}\left[\frac{c-b}{2} - \frac{r}{\pi}\right](1+u^2)^{-1}(dc + u^2db),$$

which according to the previous paragraph extends to a real analytic differential on $X^+ \cup X^-$ and hence on X.

Similarly β_2' extends to β_2 on X.

Clearly $d\beta_i = \omega$ and $\langle\beta_i, F_i\rangle = 0$ on X', and hence by continuity these equations hold on X. As for equation (3.10), note that $u(\gamma(t)) = -\cot t$ and $v(\gamma(t)) = \tan t$ so that $(\mathrm{Im}\ \gamma) \cap X' = \{q \in X' : u(q)v(q) = -1\}$. Therefore $\beta_1 = -r\pi^{-1}(1+u^2)^{-1}du$ on that locus so that $\gamma^*\beta_1 = -r\pi^{-1}dt$. Similarly $\gamma^*\beta_2 = -r\pi^{-1}dt$. Therefore (3.10) holds for β_1 and β_2.

Q.E.D.

Define f_i using β_i be means of (2.7). According to (3.10), the character Λ of Proposition 2.6 is just χ_s, so that f_i defines a smooth section s_{f_i} of $L^\times$. Let us calculate $\langle s_{f_2}, s_{f_1}\rangle$ using (2.11). We have that the ϕ of that formula must satisfy

$$(3.14)\quad d\phi = \lambda(u-v)^{-1}d(u-v) - \lambda u(1+u^2)^{-1}du - \\ - \lambda v(1+v^2)^{-1}dv + r\pi^{-1}(1+u^2)^{-1}du - r\pi^{-1}(1+v^2)dv$$

on X', that ϕ must be smooth on X, and that $\phi(p) = 0$. Up to an additive constant on each component of X' we have

$$(3.15)\quad \phi = \lambda \log \left[\frac{|u-v|}{(1+u^2)^{1/2}(1+v^2)^{1/2}}\right] + \frac{r}{\pi}[\operatorname{arc\,tan} u - \operatorname{arc\,tan} v].$$

Therefore it will suffice to choose these additive constants so that (3.15) extends continuously to X with a value of 0 at p.

Clearly the first term of (3.15) extends continuously to X (u and v cannot become simultaneously infinite on X!). However the second term does not. The correct choice of constants gives

$$(3.16)\quad \phi = \lambda \log \left[\frac{|u-v|}{(1+u^2)^{1/2}(1+v)^{1/2}}\right] + \\ + \frac{r}{\pi}\left[\arctan u - \arctan v - \frac{\pi}{2}\operatorname{sgn}(u-v)\right].$$

Indeed, the second term of (3.16) extends continuously to all of X; moreover, $u(p) = \infty$ and $v(p) = 0$ so that $\phi(p) = 0$.

Now let g_1 and g_2 be functions from X to $\mathbb{C}$ constant along the leaves of the polarizations F_1 and F_2, respectively. Let us compute the Kostant-Sternberg half-density pairing $(g_2 s_{f_2}|dv|^{1/2}, g_1, s_{f_1}|du|^{1/2})$ according to ([2], p. 152). First we calculate easily that $|dv|^{1/2}\times|\overline{du}|^{1/2} = |\lambda|^{-1/2}|u-v|$. Thus we have formally

$$(3.17) \quad (g_2 s_{f_2} |dv|^{1/2}, g_1 s_{f_1} |du|^{1/2}) =$$

$$= \iint_{u \neq v} \hat{g}_2(v)\, \overline{\hat{g}_1(u)} \left[\frac{|v-u|}{(1+u^2)^{1/2}(1+v^2)^{1/2}}\right]^{2\pi i\lambda} e^{-2ir[\arctan v - \arctan u]}$$

$$\cdot\, e^{i\pi r\, \mathrm{sgn}(v-n)} |\lambda|^{-1/2} |v-u| \cdot \frac{|\lambda|\, |du \wedge dv|}{|v-u|^2},$$

where $\hat{g}_i$ maps $\mathbb{R}$ to $\mathbb{C}$, $g_i = \hat{g}_i \circ \pi_i$ on X', π_i being the i-th coordinate projection of X' onto $\mathbb{R}$ for $i = 1,2$.

To compare this formal integral with the results of Knapp and Stein [3], we must remember that their "noncompact picture" ([3], p. 511) makes use of functions living on the group $\bar{N} = \{\bar{n}(t) : t \in \mathbb{R}\}$, where $\bar{n}(t) = \begin{bmatrix} 1 & 0 \\ -t & 1 \end{bmatrix}$ lifted to G as a one-parameter group. We choose this parametrization of $\bar{N}$ rather than the usual one $(t \mapsto \bar{n}(-t))$ because our representations are defined by right induction rather than by left induction, as is done in [3]. In addition, the Knapp-Stein setup uses as data $(-\lambda, s, F_2)$ instead of (λ, s, F_1) as we do. Of course, the bundles are identical.

Now $w = k(\pi/2)$, with $k(\cdot)$ as in (3.10), is a Weyl group representative such that $(\mathrm{ad}\, w)p = -p$. Thus the functions used by Knapp and Stein to define their representations are related to ours as follows: our functions g_1 and g_2 correspond to functions h_1 and h_2, respectively, in their set up which are defined by

$$(3.18) \quad \begin{aligned} h_1(t) &= [(g_1 \circ \pi) \cdot f_1](\bar{n}(t) w) \text{ and} \\ h_2(t) &= [(g_2 \circ \pi) \cdot f_2](\bar{n}(t)). \end{aligned}$$

Set $\mu_1(t) = (\mathrm{ad}\, \bar{n}(t))(-p)$ and $\mu_2(t) = (\mathrm{ad}\, \bar{n}(t))p$. In terms of our coordinates u and v we have

$$(3.19) \quad \mu_1(t) \leftrightarrow (t,\infty) \text{ and } \mu_2(t) \leftrightarrow (\infty,t).$$

Since $B(p,\bar{n}'(0)) = 0$, (2.7) and (3.11) give the formula

$$(3.20)\quad h_2(t) = \hat{g}_2(t)e^{2\pi i\int_0^t \mu_2^*\beta_2} = \hat{g}_2(t)(1+t^2)^{-i\pi\lambda}e^{-2ir\arctan t}.$$

The evaluation of $h_1(t)$ is only a little more complicated. In this case a good path in G to integrate over is the path $k(\cdot)$ of (3.10) from 0 to $\pi/2$ followed by the path $\bar{n}(\cdot)w$ from 0 to t. Letting γ be as in (3.10), we get

$$(3.21)\quad \begin{aligned} h_1(t) &= \hat{g}_1(t)e^{2\pi i(\int_0^{\pi/2}\gamma^*\beta_1 + \int_0^t \mu_1^*\beta_1)} = \\ &= \hat{g}_1(t)e^{-i\pi r}(1+t^2)^{i\pi\lambda}e^{-2ir\arctan t}. \end{aligned}$$

Therefore (3.17) becomes, in terms of h_1 and h_2,

$$(3.22)\quad \begin{aligned} &(g_2 s_{f_2}|dv|^{1/2},\ g_1 s_{f_1}|du|^{1/2}) = \\ &= |\lambda|^{1/2}\iint_{u\neq v} h_2(v)\overline{h_1(u)}|v-u|^{2\pi i\lambda-1}e^{-2\pi i r\,\mathrm{neg}(v-u)}|du\wedge dv|, \end{aligned}$$

where $\mathrm{neg}\ c = \frac{1}{2}(1-\mathrm{sgn}\ c)$ for $c\in\mathbb{R}$.

Thus we see that our intertwining operator is given formally by

$$(3.23)\quad (Th_2)(u) = |\lambda|^{1/2}\int_{w\neq 0}|w|^{2\pi i\lambda-1}s^{-\mathrm{neg}\ w}h_2(u+w)dw.$$

This is precisely the correct form for such intertwining operators as obtained by Knapp and Stein, apart from normalizing factors.

4. Concluding remarks.

The example we have worked out is a simple example of the half-density pairing between Hilbert spaces associated to non-Heisenberg polarizations. That is, one cannot choose local symplectic coordinates p and q so that F_1 is given by $\partial/\partial p$ and F_2 is given by

$\partial/\partial q$. Indeed, in some sense the measure of deviation from Heisenberg relatedness goes to ∞ as we go to ∞ on X, and this seems related to the fact that our kernel is singular. This phenomenon needs to be studied in general

In translating things into the "noncompact picture", we have transferred the singularity at ∞ into one along the $u = v$ diagonal. It is natural to ask to what extent the geometry of the original setting on X would give a clue to the correct way to regularize our integrals. The analytic subtleties of [3] give one pause here.

We should not fail to note that F_1 and F_2 are transverse and tractable but not completely transverse in the sense of [2]. Indeed even the pullbacks of F_1 and F_2 to the simply connected covering space of X are not completely transverse. This fact must also contribute to the singularity of the kernel we obtained.

Finally, our method will give formal intertwining operators even when no honest operator exists, for example in the cases of non-independence of polarization studied by Vergne [7] and by Rothschild and Wolf [6]. What is at stake here is probably geometric completeness of the leaves of the polarization, since the Pukanszky condition is violated.

REFERENCES

[1] Auslander, L., and Kostant, B. Invent. Math. 14, 255-354(1971).
[2] Blattner, R. J. Proc. Sympos. Pure Math. 26, 147-165(1973).
[3] Knapp, A., and Stein E. Ann. Math. (2)93, 489-578(1971).
[4] Kostant, B. Lecture Notes in Math. 170, 87-208(1970).
[5] Moore, C. Proc. Sympos. Pure Math.26, 3-44(1973).
[6] Rothschild, L., and Wolf, J. Ann. Sci. Ecole Norm. Sup.,to appear.
[7] Vergne, M. Ann. Sci. Ecole Norm. Sup. (4)3, 353-384(1970).

University of Massachusetts
Department of Mathematics and statistics
AMHERST 01002
U. S. A.

GEOMETRY AND THE METHOD OF KIRILLOV

Jonathan Brezin*

0. Introduction: Despite the fierce scrutiny to which it has been subjected since its appearance in 1962, Kirillov's "orbit picture" for the unitary dual of a connected nilpotent Lie group continues to yield new insight into the harmonic analysis of these groups. In particular, the geometry of the orbits themselves has as yet been little studied. Our intention here is to suggest a line of study in this direction.

Let N, then, denote a connected, simply-connected nilpotent Lie group, $\mathfrak{n}$ its Lie algebra, $\mathfrak{n}^*$ the vector space dual of $\mathfrak{n}$, $N^\wedge$ the unitary dual of N. Finally, let $K: \mathfrak{n}^*/\mathrm{Ad}^*N \to N^\wedge$ be the natural bijection constructed by Kirillov. A typical orbit $\Omega \in \mathfrak{n}^*/\mathrm{Ad}^*N$ is an algebraic variety in $\mathfrak{n}^*$ homeomorphic either to a single point or to a Euclidean space of even dimension $2n$. Furthermore, a basic result of L. Pukanszky (see [4], theorem 1 on p. 504) shows that Ω must contain a linear variety of dimension much larger than n -- in fact, Ω itself may be linear -- and as we shall see, the presence of large linear subvarieties in Ω imposes a certain amount of regularity on $K(\Omega)$. Some immediate evidence is given in the next section. Our main results are formulated and proved in the remaining two sections.

This note grew out of the author's study of L. Richardson's paper [6], and it is a pleasure to be able to thank Professor Richardson for a

*Research partially supported by a fellowship from the Alfred P. Sloan Foundation, to whom the author is deeply grateful.

lively correspondence concerning his work. H. Klose also was helpful in discussing the material on multiplicities in section 3. Finally, the author thanks most warmly the University of Heidelberg and, in particular, W. Beiglböck and V. Flory for their kind hospitality during the Summer of 1974 when this work was done.

1. Some immediate gratification: One can see from the work of C. C. Moore and J. Wolf [3] that linearity is a very strong condition:

1.1 Theorem. Let $\Omega \in \mathcal{N}^*/\mathrm{Ad}^*N$. Then, Ω is a linear variety if, and only if, $K(\Omega)$ defines a square-integrable representation of $N/\mathrm{kernel}\ (K(\Omega))$.

Remark: Although Moore and Wolf do not seem to mention this result explicitly, they were undoubtedly aware of it. The proof presented here is due to Roger Howe, and it was H. Klose's suggestion to look at this in the first place.

Proof. Let $\Omega^{\perp} = \{n \in \mathcal{N}: \omega(n) = 0$ for all $\omega \in \Omega\}$. Then $\Omega^{\perp}$ is an ideal in $\mathcal{N}$, and $\exp(\Omega^{\perp})$ lies in the kernel of $K(\Omega)$. We may assume, therefore, that $\Omega^{\perp} = 0$, or in other words, that the kernel of $K(\Omega)$ is discrete. Under these conditions, Moore and Wolf prove that $K(\Omega)$ is square-integrable if, and only if, Ω is a linear variety in $\mathcal{N}^*$ of codimension 1. To complete the proof, we need only observe that if Ω is a proper linear subvariety in $\mathcal{N}^*$ and $\Omega^{\perp} = 0$, then Ω must contain a basis for $\mathcal{N}^*$ and hence must have codimension 1 in $\mathcal{N}^*$. Q.E.D.

2. On Richardson's work: For the moment it is convenient to work in a rather general setting, and so we take G to be an arbitrary Lie group containing a discrete subgroup Γ with compact quotient G/Γ. (We use cosets of the form Γg.) Form $L^2(G/\Gamma)$ with respect to the unique translation-

invariant probability measure on G/Γ, and let R denote the representation of G on $L^2(G/\Gamma)$ defined by right-translation: for all $\phi \in L^2(G/\Gamma)$ and all $g \in G$, we have $[R_g\phi](x) = \phi(xg)$ for almost all $x \in G/\Gamma$.

If T is a bounded operator on $L^2(G/\Gamma)$ that commutes with all of the operators R_g, then T must map $C^\infty(G/\Gamma)$ into itself, and furthermore the linear functional $\phi \to (T\phi)(\Gamma)$ on $C^\infty(G/\Gamma)$ is a distribution on G/Γ in the sense of L. Schwartz. Let τ denote this distribution, and let $\langle\tau, \phi\rangle$ denote the value of τ on ϕ. Then, because T commutes with every R_g, we must have

$$(T\phi)(\Gamma g) = \langle\tau, R_g\phi\rangle \qquad \forall\, \phi \in C^\infty(G/\Gamma),\ g \in G.$$

Hence T is entirely determined by τ.

Suppose now that E is the orthogonal projection of $L^2(G/\Gamma)$ onto a closed R-invariant subspace $\mathcal{H}$. Then E commutes with every R_g and hence determines, as above, a distribution ε on G/Γ. The analytic properties of E are, roughly speaking, reflected in the number of derivatives ε involves. It is not really practical, for such a general $\mathcal{H}$, to make this sort of estimate on ε, even when G is actually abelian. On the other hand, $L^2(G/\Gamma)$ decomposes uniquely into a direct sum $\sum_p \oplus \mathcal{H}(p)$ of closed R-invariant subspaces, where (1) the index p runs over a countable subset $G/\Gamma^\wedge$ of the dual $G^\wedge$ of G, and (2) the restriction of R to $\mathcal{H}(p)$ has as its equivalence class a finite multiple of p. One might hope that, in view of the naturality of these "p-primary" subspaces $\mathcal{H}(p)$, one could get estimates for the corresponding distributions ε_p on G/Γ. Even this seems out of reach now, but in at least one interesting case, a reasonably complete analysis can be done, namely, when there is a closed normal subgroup H in G such that (1) $H/(H \cap \Gamma)$ is compact, and (2) p is induced from H by some character of H. We treat this case below in Theorem 2.2. First, however, we must compute ε_p, which can be done somewhat more generally:

Let us suppose, then, that H is a closed normal subgroup of G such that $H/(H \cap \Gamma)$ is compact, and let us suppose that the element p of $G/\Gamma^\wedge$ is induced from H by an element q of $(H/H \cap \Gamma)^\wedge$. Let $\mathcal{J}(p)$ denote $\{q \in (H/H \cap \Gamma)^\wedge : q$ induces $p\}$, and for each $q \in \mathcal{J}(p)$, let ε_q denote the distribution on $H/H \cap \Gamma$ corresponding to the projection of $L^2(H/H \cap \Gamma)$ onto its q-primary component. Finally, identify $\Gamma H/\Gamma$ with $H/H \cap \Gamma$ via the natural map.

2.1. Theorem. The distribution ε_p is given by the formula

$$(2.1.1) \qquad \langle \varepsilon_p, \varphi \rangle = \sum_{q \in \mathcal{J}(p)} \langle \varepsilon_q, \varphi | (\Gamma H/\Gamma) \rangle$$

for all $\varphi \in C(G/\Gamma)$.

Sketch of the proof: Let δ_p denote the distribution on G/Γ defined by the right-hand side of (2.1.1). Because $\mathcal{J}(p)$ is invariant under the usual action of Γ on $H^\wedge$, the function $\langle \delta_p, R_g\phi \rangle$ of g depends only on Γg and hence defines a function $D_p\phi$ on G/Γ. What we must prove is that D_p is the orthogonal projection of $L^2(G/\Gamma)$ onto $\mathcal{H}(p)$.

We begin by showing that $||D_p\phi||_2 \leq ||\phi||_2$. First observe that the restriction $(D_p\phi)|(\Gamma H/\Gamma)$ is simply the orthogonal projection of $\phi|(\Gamma H/\Gamma)$ into the sum of the q-primary subspaces of $L^2(H/H \cap \Gamma)$, q varying over $\mathcal{J}(p)$. Hence, for all $g \in G$ and $\phi \in C^\infty(G/\Gamma)$, we have $||(D_pR_g\phi)|(\Gamma H/\Gamma)||_2 \leq ||(R_g\phi)|(\Gamma H/\Gamma)||_2$. Integrating this last inequality over $G/\Gamma H$ yields $||D_p\phi||_2 \leq ||\phi||_2$.

Since D_p is bounded on $L^2(G/\Gamma)$ and commutes with every R_g, we must have $D_p\phi \in C^\infty(G/\Gamma)$ whenever $\phi \in C^\infty(G/\Gamma)$. Hence it makes sense to speak, for a given $g \in G$, about the family $\mathcal{K}_p(g) = \{D_p\phi|(\Gamma_g H)$: $\phi \in C^\infty(G/\Gamma)\}$ of restrictions. These functions may be viewed as being defined on $H/g^{-1}(H \cap \Gamma)g$. We can therefore form the closure $\mathcal{L}_p(g)$ of $\mathcal{K}_p(g)$ in the space $L^2(H/g^{-1}(H \cap \Gamma)g)$. Then, denoting by gq the effect of $g \in G$ on $q \in H^\wedge$, we have $\mathcal{L}_p(g) \subseteq \sum \oplus_{q \in \delta(p)} \mathcal{H}(gq)$, where $\mathcal{H}(gq)$ is the gq-primary component, as usual. Hence, if $\mathcal{L}_p$ is the closure of the image of D_p in the L^2-norm, then $(R|H)|\mathcal{L}_p$ has for its equivalence class the integral over the G-orbit in $H^\wedge$ containing $\delta(p)$. However, since p is induced from H, it is the only irreducible representation of G lying over this orbit in $H^\wedge$. Hence $\mathcal{L}_p \subseteq \mathcal{H}(p)$, and we have shown that D_p maps $L^2(G/\Gamma)$ into $\mathcal{H}(p)$.

We show next that if $\phi \in C^\infty \cap \mathcal{H}(p)$, then $D_p\phi = \phi$. (From this point on, the proof is then easily completed.) Actually, what we shall prove is that if $\phi \in C^\infty \cap \mathcal{H}(p)$, then $\phi|(\Gamma H/\Gamma)$ lies in the sum, over $q \in \delta(p)$, of the q-primary subspaces of $L^2(H/H \cap \Gamma)$:

Let $\dot{g}$ denote the double coset $\Gamma_g H$, and let $\mathcal{M}_p(\dot{g})$ denote the closure, in the $L^2(H/g^{-1}(H \cap \Gamma)g)$ - norm, of the family $[C^\infty \cap \mathcal{H}(p)]|\dot{g}$. Also, let $R(p, \dot{g})$ denote the representation of H on $\mathcal{M}_p(\dot{g})$ defined by right-translation. The irreducible summands of $R(p, \dot{g})$ form a discrete subset $s(p, \dot{g})$ of $H^\wedge$. One sees easily that $s(p, \dot{g}) = \{g \cdot q$: $q \in s(p, \dot{1})\}$. Hence $\bigcup_{g \in G} s(p, \dot{g})$ is a discrete union of G-orbits in $H^\wedge$. Since $(R|H)|\mathcal{H}(p)$ is a multiple of the integral over the single orbit of G containing $\delta^\wedge(p)$, it follows that $\bigcup s(p, \dot{g})$ must be the orbit of G containing $\delta(p)$, whence $s(p, \dot{1}) = \delta(p)$ = those elements of the orbit occuring in $L^2(H/H \cap \Gamma)$. Q.E.D.

An interesting special case occurs when the elements of $\mathfrak{S}(p)$ are characters of H-- that is when each $q \in \mathfrak{S}(p)$ is a homomorphism from H to the circle group $\mathbb{T}$ with kernel containing $\Gamma \cap H$. In this case we have for each $q \in \mathfrak{S}(p)$ the simple formula $\langle \varepsilon_q, \varphi \rangle = \int_{H/H\cap\Gamma} \varphi(x)\bar{q}(x)\,dx$, the bar denoting complex conjugation. Thus, the formula (2.1.1) becomes

$$\langle \varepsilon_p, \varphi \rangle = \sum_{q \in \mathfrak{S}(p)} \int_{H/\Gamma\cap H} \varphi(x)\,\bar{q}(x)\,dx.$$

Thus the analytic properties of ε_p are precisely those of the distribution $\sum_{q\in\mathfrak{S}(p)} \bar{q}$ on $H/H \cap \Gamma$. In effect, we have abelianized our problem. Our next result shows how this can be used:

<u>2.2. Theorem</u>. <u>Let</u> G, H, Γ, p <u>and</u> $\mathfrak{S}(p)$ <u>remain as before</u>. <u>In addition, assume</u> (1) <u>that</u> $\mathfrak{S}(p)$ <u>consists of characters</u>, <u>and</u> (2) <u>that the commutator subgroup</u> H' <u>of</u> H <u>has the property that</u> $H'/\Gamma \cap H'$ <u>is compact</u>. <u>Let</u> A <u>denote the compact abelian group</u> $H/(\Gamma \cap H)H'$. <u>Then</u> $\mathfrak{S}(p)$ <u>may be viewed as a subset of the Pontryagin dual</u> $A^\wedge$ <u>of</u> A, <u>and furthermore, the distribution</u> ε_p <u>on</u> G/Γ <u>will be a measure if, and only if, the set</u> $\mathfrak{S}(p)$ <u>belongs to the smallest ring of sets containing</u> $\{b + B\colon\ b \in A^\wedge$ <u>and</u> B <u>is a subgroup of</u> $A^\wedge\}$ -- <u>the so-called "coset ring"</u> $\mathrm{Cos}[A^\wedge]$ of $A^\wedge$.

<u>Proof:</u> This is an immediate consequence of the theorem of Rudin (in this generality) and P.J. Cohen (in general) that says (see [7], pp. 59-61) that a distribution on A whose Fourier transform is a characteristic function can be a measure if, and only if, the support of its Fourier transform is an element of $\mathrm{Cos}[A^\wedge]$. Q.E.D.

<u>Remarks</u>. (1) Special cases of (<u>2.1</u>) and (<u>2.2</u>) were first proved by

L. Richardson in [6], by entirely different methods in the case of (2.1). He assumed that G is nilpotent, connected and simply connected. Furthermore, he assumes to start that ε_p is a measure and treats only the spacial cases (i) where H has codimension 1 in G and (ii) where $\mathcal{T}(p)$ consists of characters. These assumptions make possible an entirely direct argument, first because a measure is a linear functional not only on C^∞ but also on bounded Borel functions, and second because in the nilpotent case one can choose a convenient fundamental domain -- in effect, a "unit cube" -- where there is a natural complete orthonormal system of bounded Borel functions. (2) I know of no examples in which $H'/(H' \cap \Gamma)$ is not compact. If H is nilpotent, then by classical results it is always true that $H'/(H' \cap \Gamma)$ is compact. We only assume $H'/(H' \cap \Gamma)$ is compact in order to guarantee that A is Hausdorff - we could get around this simply by taking $H^\#$ equal to the closure in H of $(\Gamma \cap H)H'$ and then setting $A = H/H^\#$. The point here is that the condition on H' is one of convenience. (3) The fundamenta problems with (2.2) are the normality of H, and the obscurity of $\mathrm{Cos}[A^\wedge]$. I know of no way of working with an H that is not normal in G. On the other hand, one can do much better than theorem (2.2) for more special groups. Our main result is theorem (2.4). Our next result is preliminary to that.

(2.3) Theorem. *Let $\mathbb{Z}$ denote the integers, and let S be an element of $\mathrm{Cos}[\mathbb{Z}^k]$. Then the Zariski closure of S in Euclidean space $\mathbb{R}^k$ is a finite union of linear varieties.*

Proof: We proceed by induction on k, the case $k = 1$ being obvious. Let $\mathrm{Cos}(\mathbb{Z}^k)$ be the set whose elements are all cosets $a + A$, where $a \in \mathbb{Z}^k$ and A is a subgroup of $\mathbb{Z}^k$, together with all the complements

$(a + A)' = \{x \in \mathbb{Z}^k: x \notin a + A\}$. Then every element of $\mathrm{Cos}[\mathbb{Z}^k]$ can be written in the form $\bigcup_{i \in F} \bigcap_{j \in J(i)} S_{ij}$, where each S'_{ij} is in $\mathrm{Cos}(\mathbb{Z}^k)$, each $J(i)$ is a finite set, and F is a finite set. Since closure commutes with finite unions, we need therefore only prove that whenever J is a finite set, and $j \to \mathcal{S}_j$ is a map from J to $\mathrm{Cos}(\mathbb{Z}^k)$, then the Zariski closure of $S = \bigcup_{j \in J} S_j$ is a finite union of linear varieties.

We can write S in the form

$$(a_1 + A_1) \cap \ldots \cap (a_m + A_m) \cap (b_1 + B_1)' \cap \ldots \cap (b_n + B_n)',$$

each $a_i + A_i$ being a coset, each $(b_j + B_j)'$ being the complement of a coset $b_j + B_j$. Now $(a_1 + A_1) \cap (a_2 + A_2)$ is either empty or of the form $c + (A_1 \cap A_2)$ for some $c \in \mathbb{Z}^k$. Hence we may assume S has the form

$$(a + A) \cap (b_1 + B_1)' \cap \ldots \cap (b_n + B_n)'.$$

The theorem is translation-invariant, and hence we may assume $a = 0$. Bearing our induction hypothesis in mind, we see that we may assume S has the form

$$A \cap (b_1 + B_1)' \cap \ldots \cap (b_n + B_n)'$$

<u>with</u> A <u>having finite index in</u> $\mathbb{Z}^k$.

Now look at B_1. If B_1 has finite index in $\mathbb{Z}^k$, then $(b_1 + B_1)'$ is a finite union of cosets of B_1. Using again that closure commutes with finite unions, we can handle each term of $(b_1 + B_1)'$ separately. Therefore we may assume that no B_j has finite index in $\mathbb{Z}^k$. Let p be any point <u>not</u> in the union of the linear varieties defined by $b_1 + B_1, \ldots, b_n + B_n$. If L is a line in $\mathbb{Q}^k$ through p, then L meets each of the sets $b_j + B_j$ in at most one point. Hence $L \cap A \cap (b_1 + B_1)' \cap \ldots \cap (b_n + B_n)'$ is Zariski dense in L. It follows that $A \cap (b_1 + B_1)' \cap \ldots \cap (b_n + B_n)'$ is dense in $\mathbb{R}^k$, and we are done. Q.E.D.

We intend to apply (2.3) in the following situation:

We suppose that G is a connected, simply connected solvable Lie group of exponential type, and let Γ and H remain, relative to G, as in the hypothesis of (2.2). Let $\Omega \in \mathfrak{g}^*/\mathrm{Ad}^*G$ satisfy $K(\Omega) = p \in G/\Gamma^\wedge$, where K is the Kirillov map. We also shall assume that H is connected, and that the Lie algebra $\mathfrak{h}$ of H polarizes some -- and hence, since H is normal is G, every -- element of Ω. Finally, we assume that there is a connected normal subgroup H_0 of G such that (1) H_0 is in the kernel of $K(\Omega)$, (2) $H_0/H_0 \cap \Gamma$ is compact, and (3) H/H_0 can be viewed as an algebraic group over $\mathbb{Q}$ in such a way that $\Gamma \cap H/\Gamma \cap H_0$ consists of rational points.

(2.4) Theorem. *With conditions as described just above, if the subset $\Omega|\mathfrak{h}$ of $\mathfrak{h}^*$ defines a connected component of an algebraic variety in $\mathfrak{h}^*$ defined over $\mathbb{Q}$, then the following three conditions are equivalent.*

(1) *The distribution* $\varepsilon_p = \varepsilon_{K(\Omega)}$ *is a measure.*

(2) *The variety defined by* $\Omega|\mathfrak{h}$ *is linear.*

(3) Ω *itself is a linear variety.*

Proof: That (1)$\Leftrightarrow$(2) is simply theorems (2.3) and (2.2). That (2)$\Leftrightarrow$(3) is Pukanszky's condition. Q.E.D.

If H is itself a nilpotent Lie group, then we can always apply this theorem, with H_0 = the trivial subgroup $\{1\}$. We also have, when G is nilpotent, the following consequence of Moore and Wolf's work:

(2.5) Theorem: If G is connected, simply connected and nilpotent, and if $\Omega \in \mathfrak{g}^*/\mathrm{Ad}^*G$ is a linear variety such that $K(\Omega)$ is in $G/\Gamma^\wedge$, then the distribution $\varepsilon_{K(\Omega)}$ is a measure.

Proof: We may assume that $K(\Omega)$ is locally faithful, and hence that the center zG of G is $\mathbb{R}$. It follows from (1.1) and [3] that the $K(\Omega)$-primary component of $L^2(G/\Gamma)$ consists of those functions on G/Γ that transform along zG according to the character $K(\Omega)|zG$. Hence if z is a real parameter describing zG, then $\varepsilon_{K(\Omega)}$ is simply integration along zG with respect to $e^{2\pi i w z}$ for some non-zero $w \in \mathbb{R}$. Q.E.D.

3. On multiplicities: We turn now to a rather different sort of problem. Let N denote, as in the introduction, a connected, simply connected, nilpotent Lie group that contains a discrete subgroup Γ with compact quotient N/Γ. We can then decompose $L^2(N/\Gamma)$ into its primary decomposition $\sum \oplus_{p \in N/\Gamma^\wedge} \mathcal{H}(p)$. Let $\mu(p)$ denote the multiplicity of p in $\mathcal{H}(p)$. Our aim is to get some idea of the global behavior of the function μ.

We begin by observing that if $\Omega \in \mathfrak{n}^*/\mathrm{Ad}^*N$, and if $t \neq 0$ is a real number, then $t\Omega = \{t\omega : \omega \in \Omega\}$ is again an element of $\mathfrak{n}^*/\mathrm{Ad}^*N$. Furthermore, if $K(\Omega)$ is in $N/\Gamma^\wedge$, then so also is $K(n\Omega)$ for every non-zero integer n. According to Moore and Wolf, if Ω is a linear variety, then $\mu(K(n\Omega))$ is a polynomial in n of degree equal to $\dim(\Omega)/2$. For more general orbits, it is not true that $\mu(K(n\Omega))$ is polynomial -- in fact, as Klose has pointed out, it is easy to produce a non-linear Ω such that $\liminf_{n\to\infty} \mu(K(n\Omega)) \leq 10$ and $\limsup_{n\to\infty} \mu(K(n\Omega)) = \infty$. In understanding this phenomenon, it would be helpful to be able to estimate

$$M(\Omega, t) = \sup_{0<n\leq t} \mu(K(n\Omega))$$

and

$$m(\Omega, t) = \inf_{t\leq n<\infty} \mu(K(n\Omega))$$

when t is large. For example, under what conditions on Ω can there exist a constant $C > 0$ such that for all $t > 0$, we have $m(\Omega, t) \leq C$? This behavior is as far as possible from the case where Ω is linear, where $m(\Omega, t)$ is asymptotic to $C\, t^{\dim(\Omega)/2}$.

We don't yet have as precise results on M and m as we would like, but the theorem below shows the direction we think such results ought to go.

For each Ω in $\mathfrak{n}^*/\mathrm{Ad}^*N$, let $\lambda^{\#}(\Omega)$ denote the dimension of the largest rational (with respect to Γ) linear variety lying in Ω. If Ω is rational, then by virtue of Pukanszky's condition, we must have $\lambda^{\#}(\Omega) - (\dim(\Omega)/2) \geq 0$. Set $\lambda(\Omega) = \lambda^{\#}(\Omega) - (\dim(\Omega)/2)$.

<u>(3.1) Theorem:</u> <u>Suppose that</u> $\Omega \in \mathfrak{n}^*/\mathrm{Ad}^*N$ <u>satisfies</u> $K(\Omega) \in N/\Gamma^\wedge$, <u>and suppose further that there exists a rational ideal</u> $\mathfrak{h}$ <u>in</u> $\mathfrak{n}$ <u>that polarizes every</u> ω <u>in</u> Ω. <u>Then there exists a positive constant</u> C (<u>depending on</u> Ω) <u>such that for every positive integer</u> n,

$$\mu(K(n\Omega)) \geq Cn^{\lambda(\Omega)} .$$

<u>Proof:</u> As usual, we may assume that $K(\Omega)$ is locally faithful, and hence that $\mathfrak{h}$ is abelian. Since $\mathfrak{h}$ is abelian, we shall identify it with its underlying Lie group. Given $n > 0$ in $\mathbb{Z}$, we shall say that $\omega \in \mathfrak{h}^*$ has denominator n if $\omega(\Gamma \cap H) \subseteq n^{-1}\mathbb{Z}$. Call ω integral if it has denominator 1.

Since $K(\Omega) \in N/\Gamma^{\wedge}$, we know that $\Omega|\mathfrak{h}$ has at least one integral point -- see [1], proposition 1. Further, if $\mathbb{Z}(\Omega|\mathfrak{h})$ is the family of all integral points in $\Omega|\mathfrak{h}$, then $\mu(K(\Omega))$ is the number of distinct Γ orbits in $\mathbb{Z}(\Omega|\mathfrak{h})$ - see [1] or [5].

Now since N/H acts, via Ad^*, on $\mathfrak{h}^*$ as a unipotent algebraic group, the orbit $\Omega|\mathfrak{h}$ is a Euclidean space, topologically, and each $\Gamma/\Gamma \cap H$ orbit is discrete. Hence there is an integral point $\omega_0 \in \Omega|\mathfrak{h}$, and there is a ball $B(\omega_0)$ about ω_0 such that if $g \in \Gamma/\Gamma \cap H$ is non-trivial, then $g(B(\omega_0) \cap \Omega|\ )$ is disjoint from $B(\omega_0) \cap \Omega|\mathfrak{h}$.

Let V be a rational linear variety in $\Omega|\mathfrak{h}$ of dimension $\lambda(\Omega)$. Then $B(\omega_0) \cap V$ contains at least $Cn^{\lambda(\Omega)}$ points with denominator n - here C is some constant depending on V. But then $n(B(\omega_0) \cap V)$ contains at least $Cn^{\lambda(\Omega)}$ integer points from $(n\Omega)|\mathfrak{h}$, and clearly no two of these points lies on the same Γ orbit. Hence $\mu(K(n\Omega)) \geq Cn^{\lambda(\Omega)}$. Q.E.D.

As a corollary, we see that $\lim_{n\to\infty} m(\Omega, n) < \infty$ is possible only if $\lambda(\Omega) = 0$. This leads one to ask whether for some constant c, we have

$$m(\Omega, n)/n^{\lambda(\Omega)} \leq C$$

for all n. Actually this may be too much to ask, and something like $\leq o(n)$ is more likely. The problem is a lack of examples with $\dim(N/H) > 1$. As for $M(\Omega, n)$, examples suggest that if $\delta(\Omega) = \dim(\Omega) - \lambda^{\#}(\Omega)$, then

$$M(\Omega, n) \leq Cn^{\lambda(\Omega) + (\delta(\Omega)/2)}.$$

However, we are far from being able to prove such a result. H. Klose has bounds for $M(\Omega, n)$ and is, as of this writing, working on some promising examples concerning both m and M. His results should appear shortly.

Bibliography

[1] Howe, R., On Frobenius reciprocity for unipotent algebraic groups over $\mathbb{Q}$, Amer. Jour. of Math. XCIII (1971) pp. 163-172.

[2] Kirillov. A. A., Unitary representations of nilpotent Lie groups. Uspekhi Matem. Nauk, 17 (1962) pp. 57-110.

[3] Moore, C. C. and Wolf, J., Square-integrable representations of nilpotent groups, Trans. Amer. Math. Soc. 185 (1973) No. 458, pp. 445-462.

[4] Pukanszky, L., On the theory of exponential groups, Trans. Amer. Math. Soc. 126 (1967) pp. 487-507.

[5] Richardson, L., Decomposition of the L^2-space of a general compact nilmanifold, Amer. Jour. of Math. XCIII (1971) pp. 173-189.

[6] Richardson, L., A class of idempotent measures on compact nilmanifolds, to appear (preprint available).

[7] Rudin, W., Fourier Analysis on Groups, Interscience, New York 1962.

University of North Carolina
Department of Mathematics
Chapel Hill, N. C. 27514.
U. S. A.

SUR LES FONCTIONS c_w DE HARISH-CHANDRA

Jacques CARMONA

Si G est un groupe de Lie semi-simple connexe réel de centre fini, les coefficients c_w sont des fonctions à valeurs opérateur qui jouent un rôle important en analyse harmonique. La théorie des intégrales d'entrelacement permet d'expliciter ces fonctions.

On fixe une décomposition d'Iwasawa $G = KAN$ d'un groupe de Lie semi-simple connexe réel G de centre fini : K est compact, A vectoriel et N unipotent. On note : M le centralisateur (respmt M' le normalisateur) de A dans K ; $\mathfrak{g}, \mathfrak{k}, \mathfrak{a}, \mathfrak{n}, \mathfrak{m}$ les algèbres de Lie respectives de G, K, A, N, M ; $\rho(H) = \frac{1}{2}\ \mathrm{Tr\ ad}\ H_{|\mathfrak{n}}$, $(H \in \mathfrak{a})$; $\mathfrak{a}^+$ la chambre de Weyl positive définie par $\mathfrak{n}$ et $A^+ = \exp(\mathfrak{a}^+)$. Tout $x \in G$ s'écrit : $x = \underline{k}(x) \exp(H(x))n(x)$, avec $\underline{k}(x) \in K$, $H(x) \in \mathfrak{a}$ et $n(x) \in N$. Le groupe de Weyl $W = M'/M$ agit sur $\mathfrak{a}$ par la représentation adjointe ; pour chaque élément s de W on fixe $s^* \in M'$ tel que $s = s^*M$, on définit $\bar{N}_s = (s^{*-1} N s^*) \cap \bar{N}$, où $\bar{N} = s_0^{*-1} N s_0^*$, ($s_0 \in W$ tel que $s_0(\mathfrak{a}^+) = -\mathfrak{a}^+$), et on choisit convenablement une mesure de Haar $d\bar{n}_s$ sur $\bar{N}_s$. On introduit une réalisation (π_γ, V_γ) (respmt (ω_ξ, H_ξ)) de chaque classe de représentation unitaire irréductible $\gamma \in \hat{K}$ (respmt $\xi \in \hat{M}$). Si dk est la mesure de Haar normalisée sur K, pour $\gamma \in \hat{K}$, $\delta \in \hat{K}$, $\nu \in \mathfrak{a}_C^*$, $x \in G$ et $T \in \mathrm{Hom}_M(V_\delta, V_\gamma)$, l'intégrale d'Eisenstein :

$$(1) \qquad E(\gamma:\delta:\nu:x)(T) = \int_K \exp(-(\nu+\rho)(H(xk)))\ \ \pi_\gamma(\underline{k}(xk))\circ T \circ \pi_\delta(k^{-1})\ dk \ ,$$

définit une fonction entière de ν ([1] ch 9). Harish-Chandra a construit des fonctions $\Phi(\gamma:\delta:.:.) : \mathfrak{a}_C^* \times A^+ \to \mathrm{Hom}_C(\mathrm{Hom}_M(V_\delta, V_\gamma))$ et $c(\gamma:\delta:w:.) : \mathfrak{a}_C^* \to \mathrm{Hom}_C(\mathrm{Hom}_M(V_\delta, V_\gamma))$, $(w \in W)$, holomorphes en ν sur un ouvert Γ complémentaire dans $\mathfrak{a}_C^*$ d'une réunion dénombrable de plans, et telles que, pour $h \in A$, $\nu \in \mathfrak{a}_C^*$ et $T \in \mathrm{Hom}_M(V_\delta, V_\gamma)$, on ait :

$$(2) \qquad \exp(\rho(\log h))\ E(\gamma:\delta:\nu:h)(T) = \sum_{w\in W} \Phi(\gamma:\delta:w\nu:h)\circ c(\gamma:\delta:w:\nu)(T) \ .$$

Si on excepte un certain nombre de cas particuliers (γ et δ triviales, [3] ; dim A = 1 , [4]), on ne sait calculer $c(\gamma:\delta:w:\nu)$ que si $w = 1$. Pour $\gamma \in \hat{K}$, $s \in W$ et $\nu \in \mathcal{A}_C^*$, la relation :

(3) $$B(\gamma:s:\nu) = \int_{\bar{N}_s} \exp(-(\nu+\rho)(H(\bar{n}_s))) \quad \pi_\gamma(\underline{k}(\bar{n}_s)^{-1}) \quad d\bar{n}_s \qquad ,$$

définit une fonction méromorphe de ν, pour $\nu \in \mathcal{A}_C^*$ et $T \in Hom_M(V_\delta, V_\gamma)$ on a :

(4) $$c(\gamma : \delta : 1 : \nu)(T) = T \circ B(\delta : s_0 : -\nu) \qquad .$$

Théorème :

Quels que soient $\gamma \in \hat{K}$, $\delta \in \hat{K}$, $s \in W$, $\nu \in \mathcal{A}_C^*$, $x \in G$, et $T \in Hom_M(V_\delta, V_\gamma)$, la relation suivante est vérifiée :

(5) $$E(\gamma : \delta : s\nu : x)\,(B(\gamma : s^{-1} : -s\,\nu)\circ T) = \ldots$$
$$\ldots = E(\gamma : \delta : \nu : x)\,(\pi_\gamma(s^*)^{-1} \circ T \circ B(\delta : s^{-1} : -s\nu) \circ \pi_\delta(s^*)) \, .$$

Si on identifie dans (2), on en déduit immédiatement, en tenant compte de (4).

Corollaire :

Quels que soient $\gamma \in \hat{K}$, $\delta \in \hat{K}$, $s \in W$, $\nu \in \mathcal{A}_C^*$ et $T \in Hom_M(V_\delta, V_\gamma)$, la relation suivante est vérifiée :

(6) $$c(\gamma : \delta : s : \nu)\,(\pi_\gamma(s^*)^{-1} \circ T \circ B(\delta : s^{-1} : -s\nu) \circ \pi_\delta(s^*)) = \ldots$$
$$\ldots = B(\gamma : s^{-1} : -s\nu) \circ T \circ B(\delta : s_0 : -s\nu) \qquad .$$

En particulier, en utilisant la régularité de $B(\delta : s : \nu)$, on retrouve, lorsque dim A = 1, l'expression usuelle :

(7) $$c(\gamma : \delta : s_0 \; : \nu)(T) = \pi_\gamma(s_0^*) \circ B(\gamma : s_0 : \bar{\nu})^* \circ T \circ \pi_\delta(s_0^*)^{-1} \qquad .$$

Pour $\gamma \in \hat{K}$ et $\xi \in \hat{M}$, on note $P(\gamma : \xi)$ le projecteur orthogonal de V_γ sur son sous-espace isotypique de type ξ, et $\hat{M}(\gamma) = \{\xi \in \hat{M} \,/\, P(\gamma:\xi) \neq 0\}$; si $T \in Hom_M(V_\delta, V_\gamma)$, l'opérateur $P(\gamma : \xi) \circ T = T \circ P(\delta : \xi)$ est nul si $\xi \notin \hat{M}(\gamma) \cap \hat{M}(\delta)$.

Lemme :

Etant donnés $\gamma \in \hat{K}$, $\delta \in \hat{K}$, $\xi \in \hat{M}(\gamma) \cap \hat{M}(\delta)$, et $T \in \mathrm{Hom}_M(V_\delta, V_\gamma)$, on peut choisir un nombre fini de couples $(A_j, B_j) \in \mathrm{Hom}_M(V_\gamma, H_\xi) \times \mathrm{Hom}_M(V_\delta, H_\xi)$ de telle sorte que :

(8) $$T_\circ P(\delta:\xi) = \sum_{1 \leq j \leq r} A_j^* {}_\circ B_j \quad .$$

Si $\xi \in \hat{M}$ et $\nu \in \mathfrak{a}_C^*$, on note $H^{\xi,\nu}$ l'espace de Hilbert isomorphe à un sous-espace fermé de $L^2(K : H_\xi)$ formé des classes de fonctions mesurables $f : G \to H_\xi$ vérifiant presque partout, pour $x \in G$, $m \in M$, $a \in A$ et $n \in N$:

(9) $$f(xman) = \exp(-(\nu + \rho)(\log a))\ \xi(m^{-1})_\circ f(x) \quad ,$$

et dont la restriction (qu'on identifie à f) à K est dans $L^2(K : H_\xi)$; l'écriture $\langle\varphi, \psi\rangle$ désigne le produit scalaire de deux éléments φ et ψ de $L^2(K : H_\xi)$. La représentation continue $\pi^{\xi,\nu}$ de G agit sur $H^{\xi,\nu}$ par :

(10) $$(\pi^{\xi,\nu}(x) f)(y) = f(x^{-1} y) \quad , (x \in G,\ y \in G,\ f \in H^{\xi,\nu}) \quad .$$

Si $H_F^{\xi,\nu}$ est est le sous-espace des vecteurs K-finis de $H^{\xi,\nu}$, pour $f \in H^{\xi,\nu}$, la relation (voir [2]) :

(11) $$(A(s : \xi : \nu)f)(x) = \int_{\bar{N}_s} f(xs^* \bar{n}_s)\, d\bar{n}_s \quad , (x \in G) \quad ,$$

permet de définir une fonction méromorphe $\mathfrak{a}_C^* \to H_F^{s\xi, s\nu}$ vérifiant :

(12) $$\pi^{s\xi, s\nu}(x)_\circ A(s : \xi : \nu)(f) = A(s:\xi:\nu)_\circ \pi^{\xi,\nu}(x)\ (f) \quad , (x \in G) \quad .$$

Si $\gamma \in \hat{K}$, on note $H_\gamma^{\xi,\nu}$ le sous-espace de $H_F^{\xi,\nu}$ isotypique de type γ pour $\pi^{\xi,\nu}|_K$; la relation ([5] ch 8) :

(13) $$(L(\gamma:\xi:\nu)(v \otimes A))(x) = \exp(-(\nu + \rho)(H(x)))\ A_\circ \pi_\gamma(\underline{k}(x))^{-1}(v) \ , (x \in G),$$

définit un K-isomorphisme $L(\gamma:\xi:\nu) : V_\gamma \otimes \mathrm{Hom}_M(V_\gamma, H_\xi) \to H_\gamma^{\xi,\nu}$ vérifiant la condition suivante :

(14) $A(s:\xi:\nu)_\circ L(\gamma:\xi:\nu)(v\otimes A) = L(\gamma:s\xi:s\nu)(v\otimes A_\circ B(\gamma:s:\nu)_\circ \pi_\gamma(s^*)^{-1})$.

L'application du Lemme 1 et ([2], [5]) de :

(15) $\langle A(s:\xi:\nu)f, g\rangle = \langle f, A(s^{-1}:s\xi:-s\bar{\nu})g\rangle$, $(f\in H_F^{\xi,\nu}$, $g \in H_F^{s\xi,-s\bar{\nu}})$,

(16) $\langle \pi^{\xi,\bar{\nu}}(x^{-1})_\circ L(\gamma:\xi:\bar{\nu})(v\otimes A), L(\delta:\xi:\nu')(w\otimes B)\rangle = \langle v, E(\gamma:\delta:\nu:x)(A^* B)w\rangle$,

à l'expression

(17) $\langle A(s:\xi:\bar{\nu})_\circ \pi^{\xi,\bar{\nu}}(x^{-1})_\circ L(\gamma:\xi:\bar{\nu})(v\otimes A), L(\gamma:s\xi:-s\nu)(w\otimes B_\circ \pi_\delta(s^*)^{-1})\rangle$,

conduit, pour $\gamma \in \hat{K}$, $\delta \in \hat{K}$, $\xi \in \hat{M}$, $x \in G$, $\nu \in \mathcal{A}_C^*$, $s \in W$ et $T \in \mathrm{Hom}_M(V_\delta, V_\gamma)$, à l'identité :

(18) $E(\gamma : \delta : s\nu:x)(\pi_\gamma(s^*)_\circ B(\gamma:s:\bar{\nu})^*_\circ T_\circ P(\delta : \xi)_\circ \pi_\delta(s^*)^{-1} = \ldots$

$\ldots = E(\gamma:\delta:\nu:x)(T_\circ P(\delta:\xi)_\circ \pi_\delta(s^*)^{-1}{}_\circ B(\delta:s^{-1}:-s\nu)_\circ \pi_\delta(s^*))$.

Il suffit alors d'utiliser la relation :

(19) $\pi_\gamma(s^*)_\circ B(\gamma:s:\bar{\nu})^* = B(\gamma:s^{-1}:-s\nu)_\circ \pi_\gamma(s^*)$.

La méthode utilisée permet d'établir quelques relations intéressantes. Si $\mathcal{U}(\mathcal{G}_C)$, $\mathcal{U}(\mathcal{A}_C)$, $\mathcal{U}(\mathcal{K}_C)$ désignent les algèbres enveloppantes respectives de $\mathcal{G}_C$, $\mathcal{A}_C$, $\mathcal{K}_C$ et si $\mathfrak{X}$ est le centralisateur de $\mathcal{K}$ dans $\mathcal{U}(\mathcal{G}_C)$, la relation :

(20) $D = \sum_{j=1}^{j=s} H_j D_j \mod \mathcal{N}\mathcal{U}(\mathcal{G}_C)$, $(H_j \in \mathcal{U}(\mathcal{A}_C)$, $D_j \in \mathcal{U}(\mathcal{K}_C)$, $1 \le j \le s)$,

permet, pour $D \in \mathfrak{X}$, $\gamma \in \hat{K}$ et $\nu \in \mathcal{A}_C^*$, de définir $\omega(\gamma:\nu:D) \in \mathrm{Hom}_M(V_\gamma, V_\gamma)$ par :

(21) $\omega(\gamma:\nu : D) = \sum_{j=1}^{j=s} < H_j, \exp(-(\nu + \rho)) > \Pi_\gamma(D)$.

On peut alors énoncer le résultat suivant.

Proposition :

Pour $\gamma \in \hat{K}$, $\xi \in \hat{M}$, $\nu \in \mathfrak{a}^*_C$, $D \in \mathfrak{X}$, $v \in V_\gamma$, $x \in G$ et $A \in \mathrm{Hom}_M(V_\gamma, H_\xi)$, la relation suivante est vérifiée :

(22) $$\pi^{\xi,\nu}(D)_\circ L(\gamma:\xi:\nu)(v \otimes A) = \ldots$$

$$\ldots = L(\gamma:\xi:\nu)(v \otimes A_\circ \omega(\gamma:\nu:D)) \quad .$$

Corollaire :

Pour $\gamma \in \hat{K}$, $\nu \in \mathfrak{a}^*_C$, $s \in W$; et $D \in \mathfrak{X}$, la relation suivante est vérifiée :

(23) $$B(\gamma:s:\nu)_\circ \Pi_\gamma(s^*)^{-1} {}_\circ\, \omega(\gamma:s\nu:D) = \omega(\gamma:\nu:D)_\circ B(\gamma:s:\nu)_\circ \Pi_\gamma(s^*)^{-1} \quad .$$

References :

[1] WARNER, G. : Harmonic Analysis on semi-simple Lie groups I. Berlin - Heidelberg-New-York, Springer, 1972.

[2] SCHIFFMAN, G. : Bull. Soc. Math. France, 99, 3-72, (1971)

[3] HARISH-CHANDRA : Amer. J. Math. 80, 241-310, (1958).

[4] HARISH-CHANDRA : Lecture Notes in Mathematics 266, 123-149 (1972).

[5] WALLACH, N. R. : Harmonic Analysis on Homogeneous spaces New-York, Marcel Dekker, 1973.

Université d'Aix-Marseille II
U. E. R. scientifique de Luminy
70, route Léon Lachamp

13288 - MARSEILLE CEDEX 2

Sur certains quotients de l'algèbre enveloppante d'une algèbre de Lie semi-simple.

Nicole CONZE-BERLINE

Soit $\underline{g}$ une algèbre de Lie semi-simple sur un corps k de caractéristique 0, algébriquement clos. Notons $U(\underline{g})$ l'algèbre enveloppante de $\underline{g}$. Nous montrons que, lorsque I varie dans l'ensemble des idéaux primitifs minimaux de $U(\underline{g})$, les corps $\mathrm{Fract}(U(\underline{g})/I)$ sont tous isomorphes au corps $\mathrm{Fract}(\mathcal{A}_n)$, avec $n=(1/2)(\dim\underline{g} - \mathrm{rang}\,\underline{g})$, où $\mathcal{A}_n$ est l'algèbre de Weyl sur k, définie par les générateurs $p_1,\ldots,p_n,q_1,\ldots,q_n$ et les relations

$$[p_i,q_j]=0 \text{ si } i\neq j\,,\ [p_i,q_i]=1\,,\ [p_i,p_j]=[q_i,q_j]=0\,,\ \text{pour } 1\leqslant i,j\leqslant n\,.$$

Nous n'exposerons ici que les principaux résultats. Ontrouvera des démonstrations complètes dans[2].

Le point de départ est la description, due à M.DUFLO, des idéaux primitifs minimaux de $U(\underline{g})$ comme annulateurs des modules de Verma, que nous allons rappeler. Fixons une sous-algèbre de Cartan $\underline{h}$ de $\underline{g}$ et un ordre sur le système des racines de $\underline{h}$ dans $\underline{g}$. Notons $\underline{n}_+$ (resp. $\underline{n}$) la sous-algèbre de $\underline{g}$, somme des sous-espaces radiciels correspondant aux racines positives (resp. négatives). On a

$$\underline{g} = \underline{n} \oplus \underline{h} \oplus \underline{n}_+$$

Soit $\underline{h}^*$ l'espace vectoriel dual de $\underline{h}$. A tout $\lambda \in \underline{h}^*$, on associe le module de Verma de plus haut poids λ, défini par

$$M_\lambda = U(\underline{g})/J_\lambda$$

où J_λ est l'idéal à gauche de $U(\underline{g})$ engendré par $\underline{n}_+$ et les éléments de la forme $H-\lambda(H)$, pour $H \in \underline{h}$. On note ρ_λ la représentation naturelle de $U(\underline{g})$ dans M_λ et I_λ son noyau. On voit facilement que le centre Z de $U(\underline{g})$ opère scalairement sur M_λ et que le caractère χ_λ de Z ainsi défini est donné par la formule

$$\chi_\lambda(z) = P(z)(\lambda) \quad , \text{ pour } z \in Z$$

où P est la projection de $U(\underline{g})$ sur $U(\underline{h})$ associée à la décomposition

$$U(\underline{g}) = U(\underline{h}) \oplus (\underline{n}\, U(\underline{g}) + U(\underline{g})\, \underline{n}_+)$$

(on identifie $U(\underline{h})$ à l'algèbre $S(\underline{h})$ des fonctions polynômes sur $\underline{h}^*$). En notant δ la demi-somme des racines positives, on a donc $\chi_\lambda = \chi_\mu$ si et seulement si $\lambda+\delta$ et $\mu+\delta$ sont conjuguées par le groupe de Weyl. On démontre, [6], que pour tout $\lambda \in \underline{h}^*$ il existe $\mu \in \underline{h}^*$ tel que $\chi_\lambda = \chi_\mu$ et tel que M_μ soit irréductible. Ceci étant, on a le résultat suivant:

Théorème 1. ([3], th. III.2). Pour tout $\lambda \in \underline{h}^*$, l'idéal I_λ est engendré par son intersection $\operatorname{Ker}\chi_\lambda$ avec Z , et est primitif. Pour tout idéal primitif minimal I de $U(\underline{g})$ il existe $\lambda \in \underline{h}^*$ tel que $I = I_\lambda$.

On note $S(\underline{g})$ l'algèbre symétrique de $\underline{g}$. Pour chaque $\lambda \in \underline{h}^*$ la symétrisation $\sigma : S(\underline{g}) \longrightarrow U(\underline{g})$ induit une bijection linéaire de $S(\underline{n})$ sur M_λ . Soit R_λ la représentation de $U(\underline{g})$ dans $S(\underline{n})$ déduite de ρ_λ à l'aide de cette bijection. Soit $\mathcal{A}$ la sous-algèbre de $\operatorname{End}_k S(\underline{n})$ formée des opérateurs différentiels à coefficients polynômes, ($\mathcal{A}$ est isomorphe à l'algèbre de Weyl $\mathcal{A}_n$, pour $n = \dim \underline{n}$). Grâce aux formules qui (pour une algèbre de Lie quelconque) permettent de calculer les éléments de la forme $\sigma^{-1}(X\,\sigma(x))$, pour $X \in \underline{g}$ et $x \in S(\underline{g})$, ([1], th. III et [5], Lemme 2.1), nous obtenons le résultat suivant:

Théorème 2. Pour tout $\lambda \in \underline{h}^*$ on a $R_\lambda(U(\underline{g})) \subset \mathcal{A}$.

En fait, pour des besoins ultérieurs, nous calculons explicitement l'opérateur $R(X)$ dans le cas où X est soit un élément de $\underline{n}$, soit un élément de $\underline{h}$, soit un élément de $\underline{n}_+$ associé à une racine simple. Mais le théorème 2 est un cas particulier du résultat suivant:

Soient $\underline{g}$ une algèbre de Lie de dimension finie sur k et $\underline{p}$ une sous-algèbre de $\underline{g}$. Supposons que $\underline{p}$ admette comme sous-espace supplémentaire dans $\underline{g}$ une sous-algèbre $\underline{m}$ telle que $\operatorname{ad}X$ soit nilpotent pour tout $X \in \underline{m}$.

Soient τ une représentation de $\underline{p}$ dans un espace F et ρ la représentation induite à $\underline{g}$ par τ. L'espace de ρ est, par définition, $M_\rho = U(\underline{g}) \otimes_{U(\underline{p})} F$. La symétrisation induit une bijection linéaire de $S(\underline{m}) \otimes F$ sur M_ρ. Soit R la représentation de $\underline{g}$ dans $S(\underline{m}) \otimes F$ déduite de ρ à l'aide de cette bijection. Soit $\mathcal{A}(\underline{m})$ la sous- algèbre de $\mathrm{End}_k S(\underline{m})$ formée des opérateurs différentiels à coefficients polynômes. On a

$$R(U(\underline{g})) \subset \mathcal{A}(\underline{m}) \otimes \tau(U(\underline{p}))$$

Le module de Verma M_λ correspond au cas où $\underline{p}$ est la sous-algèbre de Borel $\underline{h} + \underline{n}_+$ et où τ est la représentation de dimension 1 de $\underline{p}$ qui prolonge λ. Un autre cas intéressant est celui où on prend pour $\underline{p}$ une sous-algèbre parabolique et pour $\underline{m}$ la somme des sous-espaces radiciels non contenus dans $\underline{p}$.

La deuxième étape consiste à démontrer le résultat suivant:

Théorème 3. Soit $\lambda \in \underline{h}^*$ tel que M_λ soit irréductible. Le corps des fractions de $R_\lambda(U(\underline{g}))$ coïncide avec celui de $\mathcal{A}$. (Autrement dit, pour tout $A \in \mathcal{A}$, il existe S et $S' \in R_\lambda(U(\underline{g}))$ tels que $S \neq 0$ et $A.S = S'$).

Pour cela, suivant une suggestion de M. DUFLO, nous étudions le $\underline{g}$-module $L(M_\lambda, M_\mu)$ formé des k-homomorphismes $\underline{g}$-finis de M_λ dans M_μ. Les résultats obtenus sont intéressants par eux-mêmes :

Proposition. Soient $\lambda, \mu \in \underline{h}^*$ et π une représentation irréductible de dimension finie de $\underline{g}$. La multiplicité de π dans $L(M_\lambda, M_\mu)$ est supérieure ou égale à la multiplicité du poids $\mu - \lambda$ dans π. Si M_μ est irréductible il y a égalité.

Indiquons le principe de la démonstration : si $u \in J_\mu$ on a $P(u)(\mu) = 0$ ce qui permet de définir $P(m)(\mu)$ pour $m \in M_\mu$. Notant m_λ l'image de 1 dans M_λ, on associe à tout $T \in \mathrm{Hom}_k(M_\lambda, M_\mu)$ une forme linéaire φ_T sur $U(\underline{g})$ par la formule

$$\varphi_T(u) = P((u.T)m_\lambda)(\mu) \quad , \text{ pour } u \in U(\underline{g})$$

Si M_μ est irréductible, on montre que ceci définit un isomorphisme de $\underline{g}$-modules de $L(M_\lambda, M_\mu)$ sur l'espace $\mathcal{E}_{\mu-\lambda}$ des formes linéaires φ sur $U(\underline{g})$ qui sont $\underline{g}$-finies (pour l'action de $\underline{g}$ contragrédiente de la multiplication à gauche sur $U(\underline{g})$) et qui vérifient

$$\varphi(uH) = (\mu-\lambda)(H)\ \varphi(u) \quad \text{pour } u \in U(\underline{g}) \text{ et } H \in \underline{h} .$$

Il est facile de voir que la multiplicité de π dans $\mathcal{E}_{\mu-\lambda}$ est égale à la multiplicité du poids $\mu-\lambda$ dans π. Le cas général se déduit du cas où M_μ est irréductible par passage à la limite.

Lorsque $\lambda = \mu$, on en déduit le corollaire suivant, par comparaison des multiplicités dans les $\underline{g}$-modules localement finis $L(M_\lambda, M_\lambda)$ et $U(\underline{g})/I_\lambda$:

Corollaire. Si M_λ est irréductible, on a $L(M_\lambda, M_\lambda) = \rho_\lambda(U(\underline{g}))$.

Ce corollaire prouve, pour les modules de Verma, la conjecture suivante de B.KOSTANT: soit ρ une représentation irréductible de $\underline{g}$ dans un espace vectoriel V ; tout k-endomorphisme $\underline{g}$-fini de V est de la forme $\rho(u)$ pour un $u \in U(\underline{g})$. Pour tout $\lambda \in \underline{h}^*$, le module de Verma M_λ admet un unique quotient irréductible E_λ qui est, à isomorphisme près, le seul $\underline{g}$-module irréductible de plus haut poids λ. Il serait intéressant d'étudier pour E_λ la conjecture de Kostant. Ce problème est lié à l'étude des algèbres quotients de $U(\underline{g})$ par des idéaux primitifs non nécessairement minimaux.

La fin de la démonstration du théorème 3 repose sur le lemme suivant:

Lemme. Soit $\lambda \in \underline{h}^*$. Soit A un élément de $\mathcal{A}$. On considère A comme endomorphisme de M_λ. Il existe un poids ξ et un élément T non nul de $L(M_{\lambda+\xi}, M_\lambda)$ tels que $A \cdot T \in L(M_{\lambda+\xi}, M_\lambda)$

La démonstration de ce lemme, assez compliquée, est inspirée, comme d'ailleurs la Proposition ci-dessus, par un mémoire de I.M. GELFAND et A.A. KIRILLOV [4], où sont étudiées les fonctions régulières et les opérateurs différentiels sur la variété G/N_+ quotient du groupe algébrique simplement connexe G d'algèbre de Lie $\underline{g}$ par le sous-groupe N_+ correspondant à $\underline{n}_+$.

<u>Remarque</u>. A l'exception de la dernière assertion du Théorème 1 , tous les énoncés de cet article restent vrais si l'on ne suppose plus le corps de base k algébriquement clos, à condition de supposer $\underline{g}$ déployée et $\underline{h}$ déployante. On peut même, dans la première partie du théorème 1 et dans le théorème 2, remplacer k par un anneau commutatif à unité $\mathcal{L}$, et la démonstration du Lemme utilise en fait cette généralisation, en prenant pour $\mathcal{L}$ une copie de $U(\underline{h})$.

La remarque simple, mais fondamentale, qui fait le lien entre notre travail et [4] est que les modules de Verma sont des quotients du module $M_+ = U(\underline{g})/U(\underline{g})\,\underline{n}_+$ qui peut être considéré comme l'espace des distributions de support l'origine sur G/N_+ . D'ailleurs le sens profond de certaines démonstrations de [4] (en particulier le théorème 2) provient des faits suivants, découverts en collaboration avec M. DUFLO :

Notons m_+ l'image de 1 dans M_+ . Comme $\underline{h}$ stabilise $\underline{n}_+$, l'algèbre $U(\underline{h})$ opère à droite sur M_+ par

$$(u\, m_+).h = u\, h\, m_+ \quad , \text{ pour } u \in U(\underline{g}) \text{ et } h \in U(\underline{h}) .$$

On munit l'espace $\mathrm{End}_k M_+$ de la structure de $(U(\underline{g})\otimes U(\underline{g}) , U(\underline{h})\otimes U(\underline{h}))$-bimodule définie par

$$((u\otimes v).D)(m) = u\, D(v^t\, m) \quad , \quad (D.(h\otimes h'))(m) = D(m\, h')\, h^t \quad , \text{ pour}$$

$u , v \in U(\underline{g})$, $h , h' \in U(\underline{h})$, $m \in M_+$, $D \in \mathrm{End}_k M_+$, et en notant y^t l'image d'unélément y de $U(\underline{g})$ par l'antiautomorphisme principal de $U(\underline{g})$.

Soit $\underline{k}$ la sous-algèbre de l'algèbre de Lie $\underline{g}\times\underline{g}$ (produit direct de deux copies de $\underline{g}$) par l'homomorphisme diagonal. Soit $\mathcal{R}$ le sous-bimodule de $\mathrm{End}_k M_+$ formé des éléments $\underline{k}$-finis , (on identifie $U(\underline{g})\otimes U(\underline{g})$ à $U(\underline{g}\times\underline{g})$).

D'autre part, on munit l'espace $\mathrm{Hom}_k(U(\underline{g}\times\underline{g}) , U(\underline{h}))$ de la structure de $(U(\underline{g}\times\underline{g}) , U(\underline{h}\times\underline{h}))$-bimodule définie par

$(a.\phi)(b) = \phi(a^t b)$, $(\phi.\ell)(b) = \phi(b\,\ell)$ pour a , $b \in U(\underline{g}\times\underline{g})$,

$\ell \in U(\underline{h}\times\underline{h})$ et $\phi \in \mathrm{Hom}_{\underline{k}}(\, U(\underline{g}\ \underline{g})\,,\, U(\underline{h})\,)$.

Soit $\mathcal{X}$ le sous-bimodule de $\mathrm{Hom}_{\underline{k}}(\, U(\underline{g}\times\underline{g})\,,\, U(\underline{h})\,)$ formé des éléments ϕ qui satisfont aux trois conditions suivantes :

(i) ϕ s'annule sur l'idéal à gauche de $U(\underline{g}\times\underline{g})$ engendré par $\underline{n}\times\underline{n}_+$.

(ii) $\phi(u\,h \otimes v) = h^t\,\phi(u\otimes v)$, pour u , $v \in U(\underline{g})$ et $h \in U(\underline{h})$.

(iii) ϕ est $\underline{k}$-finie

Comme P s'annule sur $U(\underline{g})\underline{n}_+$, l'élément $P(m)$ de $U(\underline{h})$ est bien défini pour $m \in M_+$. Ceci permet d'associer à tout $D \in \mathrm{End}_{\underline{k}}\, M_+$ un élément $\phi_D \in \mathrm{Hom}_{\underline{k}}(\, U(\underline{g}\times\underline{g})\,,\, U(\underline{h})\,)$ défini par

$$\phi_D(u\otimes v) = P(\, u^t\, D(v\, m_+)\,) \quad , \text{ pour } u\,,\, v \in U(\underline{g})\,.$$

On montre que ceci définit un homomorphisme injectif de $(\, U(\underline{g}\times\underline{g})\,,\, U(\underline{h}\times\underline{h})\,)$-bimodules de $\mathcal{R}$ dans $\mathcal{X}$, (comparer à la Proposition). De plus , en se ramenant au cas où $\phi \in \mathcal{X}$ est annulée par $j(\underline{n}_+)$, on montre que pour toute $\phi \in \mathcal{X}$ il existe $h \in U(\underline{h})$ et $D \in \mathcal{R}$ tels que $\phi.(h\otimes 1) = \phi_D$.

De la décomposition $\underline{g}\times\underline{g} = \underline{k} \oplus (\underline{n}\times\underline{n}_+) \oplus (\underline{h}\times\{0\})$ on déduit que la restriction à $U(\underline{k})$ est un isomorphisme de $U(\underline{k})$-modules de $\mathcal{X}$ sur l'espace $\mathcal{F}$ des applications linéaires $\underline{k}$-finies (pour l'action de $U(\underline{k})$ déduite de la multiplication à gauche dans $U(\underline{k})$) de $U(\underline{k})$ dans $U(\underline{h})$. Si on identifie $\underline{h}$ à $\underline{h}_1 = (\underline{h}\times\{0\}) \subset \underline{h}\times\underline{h}$ et si on fait opérer $U(\underline{h})$ à droite sur $\mathcal{F}$ par

$$(\psi\,.\,h)(u) = \psi(u)\,h^t \quad , \text{ pour } \psi \in \mathcal{F}\ ,\ u \in U(\underline{k}) \text{ et } h \in U(\underline{h})\ ,$$

c'est aussi un morphisme de $U(\underline{h}_1)$-modules.

Or si on note $\mathcal{E}$ l'espace des formes linéaires $\underline{k}$-finies sur $U(\underline{k})$, on peut montrer que l'injection naturelle de $\mathcal{E}\otimes U(\underline{h})$ dans $\mathcal{F}$ est un isomorphisme de $\mathcal{E}\otimes U(\underline{h})$ sur $\mathcal{F}$. A partir de là, on déduit facilement la structure de $\underline{k}$-module de $\mathcal{R}$ de celle, bien connue, de $\mathcal{E}$. Une partie des résultats de [4] s'en déduit en comparant $\mathcal{R}$ à l'anneau des opérateurs différentiels sur G/N_+.

Si $k = \mathbb{C}$, l'algèbre $\underline{g} \times \underline{g}$ s'identifie à la complexifiée de l'algèbre de Lie de G considéré comme groupe de Lie réel, et $\mathcal{X}$ apparaît comme un $U(\underline{g} \times \underline{g})$-module "générique" à l'égard des modules d'Harisch-Chandra associés à la série principale de G. Il se peut que les considérations qui précèdent permettent d'étudier ces modules, en particulier leurs critères d'irréductibilité, par des moyens purement algébriques.

Bibliographie.

1 . BEREZIN (F.A.)- Quelques remarques sur les enveloppes associatives des algèbres de Lie. Fonct. Analiz i evo pril., t.1, n°2, 1967, 1-14

2 . CONZE (N.)- Algèbres d'opérateurs différentiels et quotients des algèbres enveloppantes. Bull. Soc. Math. France, 1974, (à paraître).

3 . DUFLO (M.)- Construction of primitive ideals in an enveloping algebra. Publ. of 1971 Summer School in Math., edited by I.M. GELFAND, Bolyai-Janos Math. Soc., Budapest.

4 . GELFAND (I.M.) et KIRILLOV (A.A.)- Structure du corps enveloppant d'une algèbre de Lie semi-simple. Fonct. Analiz i evo pril., t.3, n°1, 1969, 7-26.

5 . GOODMAN (R.)- Differential operators of infinite order on a Lie group, II. Indiana Univ. Math. Jour., t.21, 1971, 383-409

6 . VERMA (D.N.)- Stucture of certain induced representations on complex semi-simple Lie algebras. Ph. D. Dissertation, Yale Univ., 1966.

Université Paris VII
U. E. R. de Mathématiques
2, place Jussieu

75230 PARIS CEDEX 05

IDEAUX PRIMITIFS COMPLETEMENT PREMIERS

DANS L'ALGEBRE ENVELOPPANTE DE $\underline{sl}(3,\mathbb{C})$.

Jacques DIXMIER

Pour toute algèbre de Lie complexe $\underline{a}$, nous noterons $U(\underline{a})$ l'algèbre enveloppante de $\underline{a}$, et Prim $U(\underline{a})$ l'espace des idéaux primitifs de $U(\underline{a})$, muni de la topologie de Jacobson. Soit $\underline{a}^*$ l'espace dual de $\underline{a}$, dans lequel le groupe adjoint algébrique A de $\underline{a}$ opère par la représentation coadjointe. Si $\underline{a}$ est résoluble, il existe une application naturelle ϕ de $\underline{a}^*/A$ dans Prim $U(\underline{a})$, qui est bijective et continue [4], [10]. (Si $\underline{a}$ est nilpotente, ϕ est même un homéomorphisme [2]; on ignore si ce résultat reste vrai pour $\underline{a}$ résoluble).

Si $\underline{a} = \underline{sl}(2,\mathbb{C})$, $\underline{a}^*/A$ s'identifie à la réunion de $\mathbb{C}$ et d'un point b, adhérent au point 0 de $\mathbb{C}$. D'autre part, Prim $U(\underline{a})$ s'identifie à la réunion de $\mathbb{C}$ et de points $b_0, b_1, b_2, \ldots$, respectivement adhérents aux points 0,1,2... de $\mathbb{C}$. Les espaces $\underline{a}^*/A$ et Prim $U(\underline{a})$ sont donc très différents. Toutefois, si l'on note Primc $U(a)$ l'ensemble des idéaux primitifs complètement premiers de $U(\underline{a})$, alors Primc $U(\underline{a}) = \mathbb{C} \cup \{b_0\}$. S'appuyant sur cet exemple (et sur l'existence de l'application de Duflo pour $\underline{a}$ quelconque), la conjecture a été émise [1] qu'il existait, pour toute algèbre de Lie complexe $\underline{a}$, une bijection continue naturelle de $\underline{a}^*/A$ sur Primc $U(a)$. (Pour $\underline{a}$ résoluble, tout idéal primitif de $U(\underline{a})$ est complètement premier, donc la restriction à Primc $U(\underline{a})$ est alors sans importance).

Nous allons déterminer tous les idéaux primitifs complètement premiers de $U(\underline{a})$ pour $\underline{a} = \underline{sl}(3,\mathbb{C})$. Nous verrons que, dans ce cas, $\underline{a}^*/A$ et Primc $U(\underline{a})$ sont non homéomorphes, et que la seule bijection "naturelle" de $\underline{a}^*/A$ sur Primc $U(\underline{a})$ est non continue.

Les précisions données dans les sections 1 à 4 sont indispensables pour la compréhension du mémoire; mais elles sont en général des conséquences faciles de la théorie des représentations, ou de théorèmes de Kostant, ou de calculs directs. On a donc donné au début peu d'indications sur les démonstrations.

Notations. Pour toute algèbre de Lie complexe $\underline{a}$, $U(\underline{a})$ a été défini ci-dessus. On notera $Z(\underline{a})$ le centre de $U(\underline{a})$, $K(\underline{a})$ le corps enveloppant de $\underline{a}$, $S(\underline{a})$ l'algèbre symétrique de $\underline{a}$, $S^n(\underline{a})$ l'ensemble des éléments de $S(\underline{a})$ homogènes de degré n, $Y(\underline{a})$ l'ensemble des éléments $\underline{a}$-invariants de $S(\underline{a})$, $\sigma_{\underline{a}}$ la symétrisation de $S(\underline{a})$ dans $U(\underline{a})$. Dans tout le mémoire, on note $\underline{g}$ l'algèbre de Lie $\underline{sl}(3,\mathbb{C})$, $\underline{h}$ la sous-algèbre de Cartan de $\underline{g}$ formée des matrices diagonales de trace nulle.

Index de futures notations importantes :

1. L'algèbre de Lie $\underline{g} = \underline{sl}\ (3,\mathbb{C})$.

1.1. Notons E_{ij} la matrice $(\alpha_{kl})_{1 \leq k,l \leq 3}$ telle que $\alpha_{kl} = 0$ pour $(k,l) \neq (i,j)$, et $\alpha_{ij} = 1$. Les racines $\pm\alpha$, $\pm\beta$, $\pm\gamma$ de $\underline{g}$ relativement à $\underline{h}$ sont telles que

$$\alpha(E_{11}-E_{22}) = 2 \qquad \alpha(E_{22}-E_{33}) = -1$$

$$\gamma(E_{11}-E_{22}) = -1 \qquad \gamma(E_{22}-E_{33}) = 2$$

$$\beta = \alpha + \gamma.$$

Avec les notations habituelles, on a ensuite

$$X_\alpha = E_{12}, X_\beta = E_{13}, X_\gamma = E_{23}, X_{-\alpha} = E_{21}, X_{-\beta} = E_{31}, X_{-\gamma} = E_{32}.$$

$$H_\alpha = [X_\alpha, X_{-\alpha}] = E_{11} - E_{22},\ H_\beta = [X_\beta, X_{-\beta}] = E_{11} - E_{33},\ H_\gamma = [X_\gamma, X_{-\gamma}] = E_{22} - E_{33}$$

d'où

$$\alpha(H_\alpha) = \beta(H_\beta) = \gamma(H_\gamma) = 2, \alpha(H_\beta) = -\alpha(H_\gamma) = \beta(H_\alpha) = \beta(H_\gamma) = -\gamma(H_\alpha) = \gamma(H_\beta) = 1.$$

Le reste de la table de multiplication est donné par

$$[X_\alpha,X_\beta] = [X_\alpha, X_{-\gamma}] = [X_\beta,X_\gamma] = [X_\gamma,X_{-\alpha}] = [X_\alpha,X_{-\beta}] = [X_{-\beta},X_{-\gamma}] = 0,$$

$$[X_\alpha,X_\gamma] = X_\beta,\ [X_\alpha,X_{-\beta}] = -X_{-\gamma},\ [X_\beta,X_{-\alpha}] = -X_\gamma,\ [X_\beta,X_{-\gamma}] = X_\alpha,$$

$$[X_\gamma,X_{-\beta}] = X_{-\alpha},\ [X_{-\alpha},X_{-\gamma}] = -X_{-\beta}.$$

1.2. Nous noterons θ l'automorphisme d'ordre 2 de $\underline{g}$ tel que $\theta(X_\alpha) = X_\gamma$, $\theta(X_\beta) = -X_\beta$, $\theta(X_{-\alpha}) = X_{-\gamma}$, $\theta(X_{-\beta}) = -X_{-\beta}$, $\theta(H_\alpha) = H_\gamma$. Nous noterons $\underline{p}$ la sous-algèbre $\underline{h} + \mathbb{C}X_\alpha + \mathbb{C}X_{-\alpha} + \mathbb{C}X_\beta + \mathbb{C}X_\gamma$.

2. L'algèbre symétrique $S(\underline{g})$.

2.1. Soient ω,ω' les éléments suivants de $S(\underline{g})$:

$$\omega = 3X_{-\alpha}X_\alpha + 3X_{-\beta}X_\beta + 3X_{-\gamma}X_\gamma + H_\alpha^2 + H_\alpha H_\gamma + H_\gamma^2$$

$$\omega' = 27\,X_{-\alpha}X_{-\gamma}X_\beta + 27\,X_{-\beta}X_\alpha X_\gamma + 9X_{-\alpha}(H_\alpha+2H_\gamma)X_\alpha + 9X_{-\beta}(H_\alpha-H_\gamma)X_\beta$$
$$-9X_{-\gamma}(2H_\alpha+H_\gamma)X_\gamma + (H_\alpha+2H_\gamma)(H_\alpha-H_\gamma)(2H_\alpha+H_\gamma).$$

Alors ω et ω' sont algébriquement indépendants, et engendrent l'algèbre $Y(\underline{g})$.

2.2. Grâce à la forme de Killing, identifions $\underline{g}$ à $\underline{g}^*$, et $S(\underline{g})$ à l'algèbre des fonctions polynomiales sur $\underline{g}$. Soit G le groupe adjoint de $\underline{g}$. Les fonctions ω, ω' sont constantes sur les G-orbites. L'application $x \mapsto (\omega(x), \omega'(x))$ de $\underline{g}$ dans $\mathbb{C}^2$ définit une application ϕ de $\underline{g}/G$ sur $\mathbb{C}^2$. Soient $\underline{g}_r$ l'ensemble des éléments réguliers de $\underline{g}$, et $\mathcal{O}_1 = \underline{g}_r/G$. Alors $\phi|\mathcal{O}_1$ est une bijection de $\mathcal{O}_1$ sur $\mathbb{C}^2$. Soit $\mathcal{O}_3 \subset \underline{g}/G$ l'ensemble réduit à la seule orbite $\{0\}$. Soit $\mathcal{O}_2 = (\underline{g}/G) - \mathcal{O}_1 - \mathcal{O}_3$. On a donc la partition $\underline{g}/G = \mathcal{O}_1 \cup \mathcal{O}_2 \cup \mathcal{O}_3$. Soit $x \in \underline{g}_r$. La dimension de Gx est 6, et les conditions suivantes sont équivalentes : 1) tout élément de Gx est semi-simple; 2) Gx est fermé; 3) $(4\omega^3-\omega'^2)(x) \neq 0$. Si $(4\omega^3-\omega'^2)(x) = 0$ mais $(\omega(x),\omega'(x)) \neq (0,0)$, $(Gx)^- - (Gx)$ est une orbite de dimension 4. Si $\omega(x) = \omega'(x) = 0$, Gx est l'ensemble des éléments nilpotents réguliers ; $(Gx)^- - (Gx)$ est réunion d'une orbite de dimension 4 (l'ensemble A des éléments nilpotents non nuls et non réguliers), et de $\{0\}$ (qui est contenu dans A^-). L'ensemble $\mathcal{O}_2$ est l'ensemble des orbites de dimension 4. L'espace $\underline{g}/G$ peut être représenté par la figure 1 . (Les points de la courbe sont doublés, sauf le point de rebroussement qui est triplé).

(voir figure 1 en annexe)

2.3. Soit P l'ensemble des poids dominants radiciels, c'est-à-dire l'ensemble des $m\beta + n\alpha$, $m\beta + n\gamma$ ($m,n = 0,1,2,\ldots$, $m \geq n$). Pour $\lambda \in P$, nous noterons E_λ un $\underline{g}$ -module simple (de dimension finie) de plus grand poids λ.

Munissons $S(\underline{g})$ de la représentation adjointe. Soit $\mathcal{H}$ l'ensemble des éléments harmoniques de $S(\underline{g})$ (c'est par exemple le sous-espace vectoriel de $S(\underline{g})$ engendré par les puissances des éléments nilpotents de $\underline{g}$). L'ensemble $\mathcal{H}$ est un sous-$\underline{g}$-module de $S(\underline{g})$. Pour $\lambda \in P$, soit $S(\underline{g})_\lambda$ (resp. $\mathcal{H}_\lambda$) la somme des sous-$\underline{g}$-modules de $S(\underline{g})$ (resp. $\mathcal{H}$) isomorphes à E_λ. On a $S(\underline{g}) = \bigoplus_{\lambda \in P} S(\underline{g})_\lambda$, $\mathcal{H} = \bigoplus_{\lambda \in P} \mathcal{H}_\lambda$.

Considérons $S(\underline{g})$ comme un $Y(\underline{g})$-module. Alors les $S(\underline{g})_\lambda$ sont des sous-$Y(\underline{g})$-modules. Toute base de $\mathcal{H}$ sur $\mathbb{C}$ est une base de $S(\underline{g})$ sur $Y(\underline{g})$ de sorte que $S(\underline{g})$ s'identifie à $\mathcal{H} \otimes_{\mathbb{C}} Y(\underline{g})$. Alors $S(\underline{g})_\lambda = \mathcal{H}_\lambda \otimes_{\mathbb{C}} Y(\underline{g})$. La multiplicité de E_λ dans $\mathcal{H}_\lambda$ est égale à la multiplicité du poids 0 dans E_λ.
(Concernant 2.3, cf. [8]).

2.4. Le $\underline{g}$-module $S^1(\underline{g}) = \underline{g}$ est isomorphe à E_β. Le $\underline{g}$-module $S^2(\underline{g})$ est isomorphe à $E_0 \oplus E_\beta \oplus E_{2\beta}$. Il existe donc un sous-$\underline{g}$-module unique de $S^2(\underline{g})$ isomorphe à E_β. Nous le noterons M_β. Pour des raisons de degré, des éléments de $\underline{g} \oplus M_\beta$ linéairement indépendants sur $\mathbb{C}$ sont linéairement indépendants sur $Y(\underline{g})$. D'autre part, la multiplicité du poids 0 dans E_β est 2. Compte-tenu de 2.3, on voit que

$$S(\underline{g})_\beta = (\underline{g} \otimes Y(\underline{g})) \oplus (M_\beta \otimes Y(\underline{g})) = \underline{g}.\ Y(\underline{g}) \oplus M_\beta.\ Y(\underline{g}).$$

2.5. Il existe un sous-$\underline{g}$-module unique de $S^3(\underline{g})$ isomorphe à $E_{\beta+\alpha}$ (resp. $E_{\beta+\gamma}$). Nous le noterons $M_{\beta+\alpha}$ (resp. $M_{\beta+\gamma}$). La multiplicité du poids 0 dans $E_{\beta+\alpha}$, $E_{\beta+\gamma}$ est 1. Donc

$$S(\underline{g})_{\beta+\alpha} = M_{\beta+\alpha} \otimes Y(\underline{g}) = M_{\beta+\alpha}.\ Y(\underline{g})$$

$$S(\underline{g})_{\beta+\gamma} = M_{\beta+\gamma} \otimes Y(\underline{g}) = M_{\beta+\gamma}.\ Y(\underline{g}).$$

3. L'algèbre enveloppante $U(\underline{g})$.

3.1. Soient z, z' les éléments suivants de $U(\underline{g})$:

$$z = \sigma_{\underline{g}}(\omega) = 3\, X_{-\alpha}X_\alpha + 3X_{-\beta}X_\beta + 3X_{-\gamma}X_\gamma + H_\alpha^2 + H_\alpha H_\gamma + H_\gamma^2 + 3H_\alpha + 3H_\gamma$$

$$z' = \sigma_{\underline{g}}(\omega') = 27\, X_{-\alpha}X_{-\gamma}X_\beta + 27\, X_{-\beta}X_\alpha X_\gamma + 9X_{-\alpha}(H_\alpha+2H_\gamma)X_\alpha + 9X_{-\beta}(H_\alpha-H_\gamma)X_\beta$$
$$- 9X_{-\gamma}(2H_\alpha+H_\gamma)X_\gamma + (H_\alpha+2H_\gamma)(H_\alpha-H_\gamma)(2H_\alpha+H_\gamma) + 27X_{-\alpha}X_\alpha - 27X_{-\gamma}X_\gamma$$
$$+ 9H_\alpha^2 - 9H_\gamma^2 + 9H_\alpha - 9H_\gamma.$$

Alors z et z' sont algébriquement indépendants, et engendrent l'algèbre $Z(\underline{g})$.

3.2. Munissons $U(\underline{g})$ de la représentation adjointe. Soit $\mathcal{K} = \sigma_{\underline{g}}(\mathcal{H})$ l'ensemble des éléments harmoniques de $U(\underline{g})$. C'est un sous-$\underline{g}$-module de $U(\underline{g})$. Pour $\lambda \in P$, définissons $U(\underline{g})_\lambda$, $\mathcal{K}_\lambda$ de manière évidente. On a $U(\underline{g}) = \oplus_{\lambda \in P} U(\underline{g})_\lambda$, $\mathcal{K} = \oplus_{\lambda \in P} \mathcal{K}_\lambda$. Considérons $U(\underline{g})$ comme un $Z(\underline{g})$-module. Alors les $U(\underline{g})_\lambda$ sont des sous-$Z(\underline{g})$-modules. Toute base de $\mathcal{K}$ sur $\mathbb{C}$ est une base de $U(\underline{g})$ sur $Z(\underline{g})$, de sorte que $U(\underline{g})$ s'identifie à $\mathcal{K} \otimes_{\mathbb{C}} Z(\underline{g})$.

Alors $U(\underline{g})_\lambda = \mathcal{K}_\lambda \otimes_{\mathbb{C}} Z(\underline{g})$.

Nous poserons $\sigma_{\underline{g}}(M_\beta) = N_\beta$, $\sigma_{\underline{g}}(M_{\beta+\alpha}) = N_{\beta+\alpha}$, $\sigma_{\underline{g}}(M_{\beta+\gamma}) = N_{\beta+\gamma}$.

$$U(\underline{g})_\beta = (\underline{g} \otimes Z(\underline{g})) \oplus (N_\beta \otimes Z(\underline{g})) = g.\ Z(\underline{g}) \oplus N_\beta.\ Z(\underline{g}),$$

$$U(\underline{g})_{\beta+\alpha} = N_{\beta+\alpha} \otimes Z(\underline{g}) = N_{\beta+\alpha}.\ Z(\underline{g}),$$

$$U(\underline{g})_{\beta+\gamma} = N_{\beta+\gamma} \otimes Z(\underline{g}) = N_{\beta+\gamma}.\ Z(\underline{g}).$$

3.3. Il existe des isomorphismes du $\underline{g}$-module $\underline{g}$ sur le $\underline{g}$-module N_β, et tous ces isomorphismes sont proportionnels. L'un de ces isomorphismes transforme la base $(X_\beta, X_\gamma, \ldots)$ de $\underline{g}$ en la base $(X'_\beta, X'_\gamma, \ldots)$ de N_β donnée par les formules suivantes :

$$X'_\beta = 3X_\alpha X_\gamma + (H_\alpha - H_\gamma - \tfrac{3}{2})\, X_\beta$$

$$X'_\gamma = 3X_{-\alpha} X_\beta - (2H_\alpha + H_\gamma + \tfrac{3}{2}) X_\gamma$$

......

$$X'_{-\beta} = 3X_{-\alpha} X_{-\gamma} + X_{-\beta}(H_\alpha - H_\gamma + \tfrac{3}{2}).$$

Pour tout $\nu \in \mathbb{C}$, nous noterons $N_{\beta,\nu}$ le sous-$\underline{g}$-module de $U(\underline{g})$de base $(X'_\beta + \nu X_\beta, X'_\gamma + \nu X_\gamma, \ldots)$. Tous ses éléments sont harmoniques.

3.4. Les éléments suivants de $U(\underline{g})$

$$p = X_\alpha X_\gamma^2 - X_{-\alpha} X_\beta^2 + H_\alpha X_\beta X_\gamma$$

$$\theta(p) = X_\gamma X_\alpha^2 - X_{-\gamma} X_\beta^2 - H_\gamma X_\beta X_\alpha$$

joueront un rôle important. On a $[H,p] = (\beta+\gamma)(H)p$ pour tout $H \in \underline{h}$ et

$$[X_\alpha, p] = 2X_\alpha X_\gamma X_\beta - H_\alpha X_\beta^2 - 2X_\alpha X_\beta X_\gamma + H_\alpha X_\beta X_\beta = 0$$

$$[X_\gamma, p] = -X_\beta X_\gamma^2 + X_\gamma X_\beta X_\gamma = 0$$

donc p est un élément de poids $\beta+\gamma$ dans $N_{\beta+\gamma}$. De même, $\theta(p)$ est un élément de poids $\beta+\alpha$ dans $N_{\beta+\alpha}$.

3.5. Notons la formule suivante :

$$X_\gamma X'_\beta - X'_\gamma X_\beta = 3X_\alpha X_\gamma^2 - 3X_\beta X_\gamma + (H_\alpha - H_\gamma - \tfrac{3}{2})X_\gamma X_\beta$$
$$+ 3X_\gamma X_\beta - 3X_{-\alpha} X_\beta^2 + (2H_\alpha + H_\gamma + \tfrac{3}{2})X_\gamma X_\beta$$
$$= 3p.$$

4. L'ensemble $\mathfrak{J}$ des idéaux primitifs complètement premiers. Sa partition.

4.1. Nous noterons $\mathfrak{J}$ l'ensemble des idéaux primitifs complètement premiers de $U(\underline{g})$.

4.2. Pour tout idéal primitif I de $U(\underline{g})$, $I \cap Z(\underline{g})$ est un idéal de codimension 1 de $Z(\underline{g})$, donc z et z' sont congrus modulo I à des scalaires que nous noterons $\lambda(I), \mu(I)$. Par restriction, on a une application $I \mapsto (\lambda(I), \mu(I))$ de $\mathfrak{J}$ dans $\mathbb{C}^2$, que nous dirons canonique.

Si $(\lambda,\mu) \in \mathbb{C}^2$, $z-\lambda$ et $z-\mu$ engendrent un idéal de codimension 1 de $Z(\underline{g})$, et un idéal primitif de $U(\underline{g})$ que nous noterons $I(\lambda,\mu)$. Cet idéal est complètement premier. Nous avons donc une application $(\lambda,\mu) \mapsto I(\lambda,\mu)$ de $\mathbb{C}^2$ dans $\mathfrak{J}$.

On a $\lambda(I(\lambda,\mu)) = \lambda$, $\mu(I(\lambda,\mu)) = \mu$. L'application canonique de $\mathfrak{J}$ dans $\mathbb{C}^2$ est donc surjective. Par contre, soit $I \in \mathfrak{J}$; alors $I_{\lambda(I),\mu(I)}$ est l'idéal bilatère de $U(\underline{g})$ engendré par $I \cap Z(\underline{g})$, et est en général distinct de I. L'application canonique de $\mathfrak{J}$ dans $\mathbb{C}^2$ n'est donc pas injective.

4.3. Nous noterons $\mathfrak{J}_1$ l'ensemble des idéaux de la forme $I(\lambda,\mu)$ pour $(\lambda,\mu) \in \mathbb{C}^2$.

C'est l'ensemble des idéaux de $U(\underline{g})$ engendrés par les idéaux de codimension 1 de $Z(\underline{g})$, et c'est aussi l'ensemble des idéaux primitifs minimaux de $U(\underline{g})$. La restriction de l'application canonique à $\mathfrak{J}_1$ est une bijection de $\mathfrak{J}_1$ sur $\mathbb{C}^2$. On peut donc considérer que l'ensemble $\mathfrak{J}_1$ est parfaitement connu.

Concernant 4.2 et 4.3, cf. par exemple [5], 8.4.3 et 8.4.4.

4.4. Nous noterons $\mathfrak{J}_3$ l'ensemble d'idéaux de $U(\underline{g})$ réduit au seul idéal $\underline{g}U(\underline{g})$. Cet idéal est primitif et complètement premier. On a $\lambda(\underline{g}\,U(\underline{g})) = \mu(\underline{g}\,U(\underline{g})) = 0$. D'autre part $I_{0,0} \neq \underline{g}U(\underline{g})$ pour bien des raisons (cf. par exemple 6.4 ci-dessous). Donc $\mathfrak{J}_3 \cap \mathfrak{J}_1 = \emptyset$.

4.5. Notons que si un idéal bilatère I de $U(\underline{g})$ contient X_β, il contient le sous-$\underline{g}$-module de $U(\underline{g})$ engendré par X_β, c'est-à-dire $\underline{g}$, d'où $I = \underline{g}\,U(\underline{g})$ si $I \neq U(\underline{g})$.

4.6. Nous poserons $\mathcal{J}_2 = \mathcal{J} - (\mathcal{J}_1 \cup \mathcal{J}_3)$. On a donc la partition $\mathcal{J} = \mathcal{J}_1 \cup \mathcal{J}_2 \cup \mathcal{J}_3$, et ce mémoire est essentiellement consacré à l'étude de $\mathcal{J}_2$.

5. Quelques localisations.

5.1. Nous noterons S le sous-ensemble $\{1, X_\beta, X_\beta^2, X_\beta^3, \ldots\}$ de $U(\underline{g})$.

5.2. Lemme. L'ensemble des éléments de la forme $u X_\beta^{-n}$, où $u \in U(\underline{g})$ (resp. $u \in U(\underline{p})$) et $n = 0,1,2,\ldots$, est une sous-algèbre de $K(\underline{g})$ (resp. $K(\underline{p})$). C'est aussi l'ensemble des éléments de la forme $X_\beta^{-n} u$, où $u \in U(\underline{g})$ (resp. $u \in U(\underline{p})$) et $n = 0,1,2,\ldots$

Soit F l'ensemble des $u X_\beta^{-n}$, où $u \in U(\underline{g})$ et $n = 0,1,2,\ldots$ Il est clair que F est un sous-espace vectoriel de $K(\underline{g})$. On a

$$X_\beta^{-1} X_\alpha = X_\alpha X_\beta^{-1} \qquad\qquad X_\beta^{-1} X_\gamma = X_\gamma X_\beta^{-1}$$

$$X_\beta^{-1} H_\alpha = H_\alpha X_\beta^{-1} + X_\beta^{-1} X_\beta X_\beta^{-1} = (H_\alpha + 1)\, X_\beta^{-1}$$

$$X_\beta^{-1} H_\gamma = H_\gamma X_\beta^{-1} + X_\beta^{-1} X_\beta X_\beta^{-1} = (H_\gamma + 1)\, X_\beta^{-1}$$

$$X_\beta^{-1} X_{-\alpha} = X_{-\alpha} X_\beta^{-1} + X_\beta^{-1} X_\gamma X_\beta^{-1} = X_{-\alpha} X_\beta^{-1} + X_\gamma X_\beta^{-2}$$

$$X_\beta^{-1} X_{-\gamma} = X_{-\gamma} X_\beta^{-1} - X_\beta^{-1} X_\alpha X_\beta^{-1} = X_{-\gamma} X_\beta^{-1} - X_\alpha X_\beta^{-2}$$

$$X_\beta^{-1} X_{-\beta} = X_{-\beta} X_\beta^{-1} - X_\beta^{-1} H_\beta X_\beta^{-1} = X_{-\beta} X_\beta^{-1} - (H_\alpha + H_\gamma + 2)\, X_\beta^{-2} .$$

Donc $X_\beta^{-1}\underline{g} \subset \Sigma_{n \geq 1}\, \underline{g}\, X_\beta^{-n}$. Pour $p = 1,2,\ldots$, on en déduit par récurrence sur p que

$$X_\beta^{-p}\, \underline{g} \subset \Sigma_{n \geq 1}\, \underline{g}\, X_\beta^{-n} .$$

Donc $X_\beta^{-p}\, U(\underline{g}) \subset F$ par récurrence sur la filtration des éléments de $U(\underline{g})$.

Il est alors clair que $F.F \subset F$, donc F est la sous-algèbre de $K(\underline{g})$ engendrée par $U(\underline{g})$ et X_β^{-1}. Un calcul analogue prouve que cette sous-algèbre est aussi l'ensemble des $X_\beta^{-n} u$, où $u \in U(\underline{g})$ et $n = 0,1,2,\ldots$ Enfin on raisonne de la même manière pour $\underline{p}$ au lieu de $\underline{g}$.

5.3. D'après 5.2 et [5], 3.6.7, S permet un calcul des fractions dans $U(\underline{g})$ et $U(\underline{p})$, et les algèbres $F = U(\underline{g})_S$, $F' = U(\underline{p})_S$ s'identifient aux sous-algèbres de $K(\underline{g})$, $K(\underline{p})$ considérées en 5.2. Les notations F, F' seront conservées dans la suite.

5.4. Lemme . On a

$$\text{(i)} \quad \frac{1}{3} z = X_{-\beta}X_\beta + X_{-\gamma}X_\gamma + u$$

$$\text{(ii)} \quad \frac{1}{9} z' + \frac{1}{2} z = X_{-\beta}X'_\beta + X_{-\gamma}X'_\gamma + u'$$

$$\text{(iii)} \quad \frac{1}{3} z X'_\beta - (\frac{1}{9} z' + \frac{1}{2} z) X_\beta = 3X_{-\gamma}p + u''$$

avec $u, u', u'' \in U(\underline{p})$.

C'est clair pour (i). D'autre part, modulo $U(\underline{p})$, on a dans $U(\underline{g})$

$$z' \equiv 27\, X_{-\gamma}X_{-\alpha}X_\beta - 27X_{-\beta}X_\beta + 27\, X_{-\beta}X_\alpha X_\gamma + 9X_{-\beta}(H_\alpha - H_\gamma)\, X_\beta$$

$$- 9X_{-\gamma}(2H_\alpha + H_\gamma)\, X_\gamma - 27\, X_{-\gamma}X_\gamma$$

$$\equiv 9X_{-\beta}(3X_\alpha X_\gamma + (H_\alpha - H_\gamma - 3)X_\beta) + 9X_{-\gamma}(3X_{-\alpha}X_\beta - (2H_\alpha + H_\gamma + 3)\, X_\gamma))$$

d'où (ii). Comme $[X_\beta, X'_\beta] = 0$, (i) et (ii) entraînent

$$\frac{1}{3} z X'_\beta - (\frac{1}{9} z' + \frac{1}{2} z)\, X_\beta \equiv X_{-\gamma}(X_\gamma X'_\beta - X'_\gamma X_\beta)$$

d'où (iii) compte tenu de 3.5.

5.5. Lemme. Soit L l'ensemble des éléments de $K(\underline{g})$ de la forme uv^{-1} où $u \in U(\underline{g})$, $v \in U(\underline{p}) - \{o\}$. Alors L est une sous-algèbre de $K(\underline{g})$. C'est aussi l'ensemble des éléments de $K(\underline{g})$ de la forme $v^{-1}u$, où $u \in U(\underline{g})$, $v \in U(\underline{p}) - \{o\}$. On a $L = K(\underline{p})\,[z,z']$ (rappelons que $[z,z'] = [z,K(\underline{p})] = [z',K(\underline{p})] = 0$).

Soit G la sous-algèbre de $K(\underline{g})$ engendrée par $K(\underline{p})$, z,z'. Evidemment, $G \subset L$. D'après 5.4 (iii), on a $X_{-\gamma} \in G$. Alors, d'après 5.4 (i), on a $X_{-\beta} \in G$. Donc $L \subset G$ et par suite $L = G$. On vérifie de même que G est l'ensemble des $v^{-1}u$ où $u \in U(\underline{g})$, $v \in U(\underline{p}) - \{o\}$.

5.6. D'après 5.5 et [5], 3.6.7, $U(\underline{p})-\{o\}$ permet un calcul des fractions dans $U(\underline{g})$, et l'algèbre $U(\underline{g})_{U(\underline{p})-\{o\}}$ s'identifie à l'algèbre L considérée en 5.5.
La notation L sera conservée dans la suite.

5.7. Lemme. Dans F' (cf.5.3), on considère les éléments suivants :

$$p_1 = X_\alpha \qquad q_1 = X_\gamma X_\beta^{-1}$$

$$p_2 = H_\alpha X_\beta^{-1} + 2X_\alpha X_\gamma X_\beta^{-2} \qquad q_2 = X_\beta$$

$$p \ \ (\text{cf.}3.4) \qquad q = \frac{1}{3} X'_\beta X_\beta^{-1} = X_\alpha X_\gamma X_\beta^{-1} + \frac{1}{3}(H_\alpha - H_\gamma - \frac{3}{2}).$$

(i) L'ensemble $\{p_1,q_1,p_2,q_2, q_2^{-1}, p,q\}$ engendre l'algèbre F'.

(ii) Les ensembles $\{p_1,q_1\},\{p_2,q_2\},\{p,q\}$ commutent deux à deux. On a

$$[p_1,q_1] = 1 \qquad [p_2,q_2] = 1 \qquad [p,q] = p.$$

(iii) La sous-algèbre A de F' engendrée par p_1,q_1,p_2,q_2,q_2^{-1} est isomorphe à l'algèbre de Weyl localisée $(A_2)_{q_2}$.

(iv) La sous-algèbre B de F' engendrée par p,q est isomorphe à l'algèbre enveloppante d'une algèbre de Lie résoluble non commutative de dimension 2

(v) On a $F' = B \otimes_{\mathbb{C}} A$.

L'assertion (i) est claire. D'après 3.3 et 3.4, p et q commutent à $H_\alpha,X_\alpha,X_\beta,X_\gamma$, donc à p_1,q_1,p_2,q_2. On a

$$[p_1,q_1] = X_\beta X_\beta^{-1} = 1 \qquad [p_2,q_2] = X_\beta X_\beta^{-1} = 1$$

$$[p_1,q_2] = 0 \qquad [q_1,q_2] = 0$$

$$[p_1,p_2] = -2X_\alpha X_\beta^{-1} + 2X_\alpha X_\beta X_\beta^{-2} = 0$$

$$[q_1,p_2] = 2X_\gamma X_\beta^{-1} X_\beta^{-1} - 2X_\beta X_\beta^{-1} X_\gamma X_\beta^{-2} = 0.$$

Puisque p commute à $X_\alpha,X_\beta,X_\gamma$, on a

$$[p,q] = [p, \frac{1}{3}(H_\alpha - H_\gamma)] = \frac{1}{3}(\beta+\gamma)(H_\gamma - H_\alpha)p = p$$

d'où (ii).

L'assertion(iii) est alors immédiate. On sait que, dans l'algèbre universelle définie par les générateurs p, q et la relation $[p,q] = p$, tout idéal non nul contient une puissance de p. Comme les puissances de p dans F' sont non nulles, on a (iv). Enfin, (v) résulte de ce qui précède puisque A est une algèbre simple centrale ([7], p.110).

5.8. Lemme. On conserve les notations de 5.7. Les idéaux premiers de F' sont :

0

$Bp \otimes A = F'p = pF'$

et, pour tout $\lambda \in \mathbb{C}$,

$(Bp + B(q-\lambda)) \otimes A = F'p + F'(q-\lambda) = pF' + (q-\lambda)F'$.

Ces idéaux sont complètement premiers.

Comme A est simple centrale, les idéaux bilatères de F' sont les ensembles $J \otimes A$, où J est un idéal bilatère de B. D'autre part $J \otimes A$ est premier (resp. complètement premier) si et seulement si J est premier (resp. complètement premier). Il suffit alors d'utiliser la liste connue (et facile) des idéaux premiers de B.

6. Etude de $\mathcal{J}_2$.

6.1. Lemme. Soient $\underline{a}$ une algèbre de Lie semi-simple complexe, I un idéal bilatère primitif de $U(\underline{a})$, $I' = I \cap Z(\underline{a})$. Les conditions suivantes sont équivalentes :

(i) I est l'idéal engendré par I';

(ii) I ne contient aucun élément harmonique non nul de $U(\underline{a})$.

Soit $\mathcal{L}$ l'ensemble des éléments harmoniques de $U(\underline{a})$. On a $U(\underline{a}) = \mathcal{L} \otimes Z(\underline{a})$, $Z(\underline{a}) = \mathbb{C}.1 \oplus I'$, donc $U(\underline{a}) = \mathcal{L} \oplus \mathcal{L}.I'$, et $U(\underline{a}).I = \mathcal{L}.Z(\underline{a}).I' = \mathcal{L}.I'$. Par suite, $I = (\mathcal{L} \cap I) \oplus (U(\underline{a}).I')$, d'où le lemme.

6.2. Lemme. Soit I un idéal complètement premier de $U(\underline{g})$ tel que $I \cap U(\underline{p}) \neq 0$. On a $I \supset N_{\beta+\gamma}$.

Si $X_\beta \in I$, on a $I = \underline{g}\, U(\underline{g})$ (4.5) donc $I \supset N_{\beta+\gamma}$. Supposons désormais $X_\beta \notin I$. Alors $X_\beta^n \notin I$ pour $n = 0,1,2,\dots$, donc $I = I_S \cap U(\underline{g})$ ([5], 3.6.17). Or $I_S \cap U(\underline{p})_S$ est un idéal non nul complètement premier de $U(\underline{p})_S$. D'après 5.8, on a $p \in I_S \cap U(\underline{p})_S$, d'où $p \in I_S \cap U(\underline{g}) = I$, et par suite $N_{\beta+\gamma} \subset I$.

6.3. Lemme. Soit I un idéal primitif de $U(\underline{g})$ tel que $I \cap U(\underline{p}) = 0$. Alors $I \in \mathcal{J}_1$.

Soit I' l'idéal de $U(\underline{g})$ engendré par $I \cap Z(\underline{g})$. On a $I' \in \mathcal{J}_1$, $I \supset I'$, et $I' \cap U(\underline{p}) = 0$. Posons $T = U(\underline{p}) - \{o\}$. On peut former I_T, I'_T qui sont des idéaux de $L = K(\underline{p})[z,z']$ distincts de L.

Puisque I' est primitif dans $U(\underline{g})$, on a $z \in \mathbb{C} + I' \subset \mathbb{C} + I'_T$, $z' \in \mathbb{C} + I' \subset \mathbb{C} + I'_T$. Par suite, $L = K(\underline{p}) + I'_T \subset K(\underline{p}) + I_T$. Mais $I_T \cap K(\underline{p}) = 0$ puisque $K(\underline{p})$ est un corps. Donc $L = K(\underline{p}) \oplus I'_T = K(\underline{p}) \oplus I_T$, d'où $I'_T = I_T$, $I = I' \in \mathcal{J}_1$.

6.4. Théorème. Soit I un idéal primitif complètement premier de $U(\underline{g})$. Les conditions suivantes sont équivalentes :

(i) $I \in \mathcal{J}_2 \cup \mathcal{J}_3$.

(ii) I contient un élément harmonique non nul.

(iii) Il existe une sous-algèbre parabolique $\underline{q} \neq \underline{g}$ de $\underline{g}$ telle que $I \cap U(\underline{q}) \neq 0$.

(iv) Pour toute sous-algèbre de Borel $\underline{b}$ de $\underline{g}$, on a $I \cap U(\underline{b}) \neq 0$.

(v) $I \supset N_{\beta+\alpha}$.

(vi) $I \supset N_{\beta+\gamma}$.

(vii) $I = \underline{g}\, U(\underline{g})$, ou bien il existe $\nu \in \mathbb{C}$ tel que $I \supset N_{\beta,\nu}$.

On a (i) $\Leftrightarrow$ (ii) d'après 6.1.

Prouvons que (i) $\Rightarrow$ (iii) $\Rightarrow$ (v) ou (vi) $\Rightarrow$ (ii).

L'implication (i) $\Rightarrow$ (iii) résulte de 6.3. Supposons qu'il existe une sous-algèbre parabolique $\underline{q} \neq \underline{g}$ telle que $I \cap U(\underline{q}) \neq 0$. En agrandissant au besoin $\underline{q}$, on peut supposer $\eta(\underline{q}) = \underline{p}$ où $\eta \in \mathrm{Aut}(\underline{g})$. Alors $\eta(I) \cap U(\underline{p}) \neq 0$. On a $\eta \in \mathrm{Int}(\underline{g})$ ou $\theta \circ \eta \in \mathrm{Int}(\underline{g})$. Comme tout idéal bilatère de $U(\underline{g})$ est invariant par automorphisme intérieur, on a $I \cap U(\underline{p}) \neq 0$ ou $\theta(I) \cap U(\underline{p}) \neq 0$. D'après 6.2, $I \supset N_{\beta+\gamma}$ ou $I \supset N_{\beta+\alpha}$. Enfin, l'implication (v) ou (vi) $\Rightarrow$ (ii) est évidente. Si $I \supset N_{\beta+\alpha}$, on a $I \notin \mathcal{J}_1$ d'après ce qui précède, donc $I \cap U(\underline{p}) \neq 0$ (6.3) et $I \supset N_{\beta+\gamma}$ (6.2). En utilisant θ, on en déduit que (vi) $\Rightarrow$ (v), donc que (v) $\Leftrightarrow$ (vi). Ainsi, les conditions (i), (ii), (iii), (v), (vi) sont équivalentes.

(vii) $\Rightarrow$ (iv) : supposons vérifiée la condition (vii). Alors, si $\underline{b}_o = \underline{h} + \mathbb{C}X_\alpha + \mathbb{C}X_\beta + \mathbb{C}X_\gamma$, il est clair que $I \cap U(\underline{b}_o) \neq 0$. Si $\underline{b}$ est une sous-algèbre de Borel de $\underline{g}$, il existe $\eta \in \mathrm{Int}(\underline{g})$ tel que $\eta(\underline{b}) = \underline{b}_o$, d'où $I \cap U(\underline{b}) \neq 0$.

(iv) $\Rightarrow$ (iii) : évident.

Enfin, supposons que I vérifie les conditions (i), (ii), (iii), (v), (vi) (lesquelles sont invariantes par Aut ($\underline{g}$) sous la forme (i)), et prouvons (vii). On peut supposer $X_\beta \notin I$ (4.5). Par automorphisme, on peut supposer $I \cap U(\underline{p}) \neq 0$.
On a $p \in I$ d'après 6.2. Soit ρ l'homomorphisme canonique de $F = U(\underline{g})_S$ sur F/I_S. Puisque $\rho(p) = 0$, $\rho(q)$ commute à $\rho(U(\underline{p}))$ d'après 5.7, et en particulier à $\rho(X_{-\alpha})$.
D'autre part ,

$$\begin{aligned}
[X_{-\gamma},q] &= - X_\alpha H_\gamma X_\beta^{-1} + X_\alpha X_\gamma X_\beta^{-1} X_\alpha X_\beta^{-1} - X_{-\gamma} \\
&= - H_\gamma X_\alpha X_\beta^{-1} - X_\alpha X_\beta^{-1} + X_\gamma X_\alpha^2 X_\beta^{-2} + X_\alpha X_\beta^{-1} - X_{-\gamma} \\
&= (X_\gamma X_\alpha^2 - X_{-\gamma} X_\beta^2 - H_\gamma X_\beta X_\alpha)\, X_\beta^{-2} \\
&= \theta\,(p)\, X_\beta^{-2} .
\end{aligned}$$

Or $\theta(p) \in N_{\beta+\alpha} \subset I \subset I_S$. Donc $\rho(q)$ commute à $\rho(X_{-\gamma})$. Ainsi, $\rho(q)$ appartient au centre de $U(\underline{g})_S/I_S$.
Comme I est primitif, le centre de Fract $(U(\underline{g})/I)$ est réduit à $\mathbb{C}$ ([5], 4.1.6 et 2.6.4). Donc il existe $\nu \in \mathbb{C}$ tel que $\rho(q) = \nu$, d'où $\rho\,(\frac{1}{3} X'_\beta) = \nu\, \rho(X_\beta)$ et $X'_\beta - 3\,\nu\, X_\beta \in I_S \cap U(\underline{g}) = I$. Alors, $N_{\beta,-3\nu} \subset I$.

6.5. <u>Théorème.</u> <u>Soit $I \in \mathfrak{J}_2$. Il existe un nombre complexe $\nu(I)$ et un seul tel que $I \supset N_{\beta,\nu(I)}$.</u>

L'existence de $\nu(I)$ résulte de 6.4 (vii). Si $N_{\beta,\xi} \subset I$ et $N_{\beta,\xi'} \subset I$, on a $X'_\beta + \xi X_\beta, X'_\beta + \xi' X_\beta \in I$, donc $(\xi'-\xi)X_\beta \in I$, donc $\xi = \xi'$ puisque $I \neq \underline{g}\, U(\underline{g})$ (cf.4.5).

6.6. <u>Lemme.</u> <u>Soit $\xi \in \mathbb{C}$. Soit ϕ la forme linéaire sur $\underline{p}$ qui s'annule sur $[p,p] = \mathbb{C}X_\alpha + \mathbb{C}H_\alpha + \mathbb{C}X_{-\alpha} + \mathbb{C}X_\beta + \mathbb{C}X_\gamma$, et telle que $\phi(H_\gamma) = \xi$. Soient ρ la représentation de $U(\underline{g})$ induite par ϕ, et $I = \mathrm{Ker}\,\rho$. On a $I \in \mathfrak{J}_2$, $\nu(I) = \xi + \frac{3}{2}$, $\lambda(I) = \xi(\xi+3)$, $\mu(I) = -\,\xi(\xi+3)(2\xi+3)$.</u>

D'après [3], I est complètement premier.
Comme $\underline{g} = \underline{p} \oplus (\mathbb{C}X_{-\beta} + \mathbb{C}X_{-\gamma})$, ρ peut se réaliser dans $U(\mathbb{C}X_{-\beta} + \mathbb{C}X_{-\gamma}) = S(\mathbb{C}X_{-\beta} + \mathbb{C}X_{-\gamma})$. On a, pour tout $H \in \underline{h}$,

$$H\, X^m_{-\beta} X^n_{-\gamma} = X^m_{-\beta} X^n_{-\gamma}\, H - (m\beta + n\gamma)\,(H)\, X^m_{-\beta}\, X^n_{-\gamma}$$

donc

$$\rho(H)(X^m_{-\beta}\, X^n_{-\gamma}) = (\phi - m\beta - n\gamma)\,(H)\, X^m_{-\beta}\, X^n_{-\gamma}.$$

Des raisonnements classiques (cf.par exemple [5],7.1.8) prouvent alors que le commutant de ρ ($U(\underline{g})$) se réduit à $\mathbb{C}$. En particulier, le centre de $U(\underline{g})/I$ est $\mathbb{C}$. D'après [6], cor.1, I est primitif. Bref, $I \in \mathfrak{J}$.

Comme $\rho\,(X_{-\beta}).\,1 = X_{-\beta}$, on a $X_{-\beta} \notin I$, donc $I \notin \mathfrak{J}_3$. D'autre part,

$$X'_{-\beta}\, X^m_{-\beta}\, X^n_{-\gamma} = 3\, X^m_{-\beta}\, X_{-\alpha}\, X^{n+1}_{-\gamma} + X^{m+1}_{-\beta}(H_\alpha + \tfrac{3}{2} - H_\gamma)\, X^n_{-\gamma}$$

$$= 3\, X^m_{-\beta}\, X^{n+1}_{-\gamma}\, X_{-\alpha} - 3X^m_{-\beta}(n+1)X_{-\beta}X^n_{-\gamma} + X^{m+1}_{-\beta}(H_\alpha - H_\gamma + \frac{3}{2}) + 3nX^{m+1}_{-\beta}\, X^n_{-\gamma}$$

donc

$$\rho(X'_{-\beta})X^m_{-\beta}\, X^n_{-\gamma} = -3(n+1)\, X^{m+1}_{-\beta}X^n_{-\gamma} - \xi\, X^{m+1}_{-\beta}\, X^n_{-\gamma} + 3n\, X^{m+1}_{-\beta}X^n_{-\gamma} + \frac{3}{2}\, X^{m+1}_{-\beta}\, X^n_{-\gamma}$$

$$= -(\xi + \tfrac{3}{2})\, X^{m+1}_{-\beta}\, X^n_{-\gamma}$$

$$= -(\xi + \tfrac{3}{2})\, \rho(X_{-\beta})\, X^m_{-\beta}\, X^n_{-\gamma}\,.$$

On en déduit que $X'_{-\beta} + (\xi + \frac{3}{2})\, X_{-\beta} \in I$, d'où $I \in \mathfrak{J}_2$ et $\nu(I) = +(\xi + \frac{3}{2})$. D'autre part, d'après 3.1, on a

$$\rho(z).1 = \rho(H^2_\gamma + 3H_\gamma).1 = (\xi^2 + 3\xi).1$$

$$\rho(z').1 = \rho(-2H^3_\gamma - 9H^2_\gamma - 9H_\gamma).1 = (-2\xi^3 - 9\xi^2 - 9\xi).\,1$$

d'où $\lambda(I) = \xi(\xi+3)$ et $\mu(I) = -\xi(\xi+3)(2\xi+3)$

(Les $\underline{g}$-modules considérés dans cette démonstration sont des quotients de modules de Verma).

6.7. <u>Lemme</u> . <u>Soit K l'intersection des éléments de $\mathfrak{J}_2$. Soit $u \mapsto \tilde{u}$ l'application canonique de $U(\underline{g})$ sur $U(\underline{g})/K$. Soit $K' = K \cap U(\underline{p})$.</u>

(i) L'application composée des applications canoniques $U(\underline{p})_S \to U(\underline{g})_S \to U(\underline{g})_S/K_S$ est surjective, de sorte que $U(\underline{g})_S/K_S = (U(\underline{g})/K)_{S^\sim} = U(\underline{p})_S/K'_S = (U(\underline{p})/K')_{S^\sim}$.

(ii) Identifions $U(\underline{p})_S$ à $B \otimes A$ avec les notations de 5.7. Alors K'_S est l'idéal de $U(\underline{p})_S$ engendré par p.

(iii) $(U(\underline{g})/K)_{S^\sim}$ s'identifie à $\mathbb{C}[q] \otimes (A_2)_{q_2}$

(iv) Soient $I \in \mathcal{J}_2$ et $\xi = \nu(I)$. Alors $(I/K)_{S^\sim} = \mathbb{C}[q]\,(3q+\xi) \otimes (A_2)_{q_2}$.

Soient $u \in U(\underline{g})$ et $n \in \mathbb{N}$ tels que $u X_\beta^n \in K$ (resp. $X_\beta^n u \in K$). Alors, pour tout $I \in \mathcal{J}_2$, on a $u X_\beta^n \in I$ (resp. $X_\beta^n u \in I$), donc $u \in I$; par suite, $u \in K$. On est donc dans les conditions d'application de [5],3.6.15. L'algèbre $(U(\underline{g})/K)_{S^\sim}$ s'identifie à $U(\underline{g})_S/K_S$.

On a

$$X_{-\alpha} X_\beta^2 \equiv X_\alpha X_\gamma^2 + H_\alpha X_\beta X_\gamma \text{ mod. } K \quad (6.4(vi))$$

$$X_{-\gamma} X_\beta^2 \equiv X_\gamma X_\alpha^2 - H_\gamma X_\beta X_\alpha \text{ mod. } K \quad (6.4(v))$$

donc $(U(\underline{p})_S)^\sim$ contient $X_{-\alpha}^\sim$ et $X_{-\gamma}^\sim$. Cela prouve (i).

Pour tout $I \in \mathcal{J}_2$, on a $p \in I \cap U(\underline{p})$. Donc $p \in K \cap U(\underline{p}) = K' \subset K'_S$. Par suite, dans $U(\underline{p})_S = B \otimes A$, K'_S est un idéal de la forme $J \otimes A$, où J est un idéal bilatère de B contenant p.

Soient $I \in \mathcal{J}_2$ et $\xi = \nu(I)$. Alors I/K est un idéal de $U(\underline{g})/K$ ne rencontrant pas $S^\sim$. Donc $(I/K)_{S^\sim}$ s'identifie à un idéal de $(U(\underline{g})/K)_{S^\sim} = U(\underline{p})_S/K'_S$. Ainsi, il existe un idéal bilatère Q de $U(\underline{p})_S$ contenant p tel que $(I/K)_{S^\sim} = Q/K'_S$.

On a $X'_\beta + \xi X_\beta \in I$, donc $3q + \xi \in I_S$. Si l'on écrit Q sous la forme $Q_1 \otimes A$ où Q_1 est un idéal de B, on a $p \in Q_1$ et $3q + \xi \in Q_1$, donc $Q_1 = Bp + B(3q+\xi)$, donc $Q = (Bp+B(3q+\xi)) \otimes A$ et finalement

(1) $(I/K)_{S^\sim} = Q/K'_S = ((Bp + B(3q+\xi))/J) \otimes A$.

Quand I parcourt $\mathcal{J}_2$, $\nu(I)$ prend toutes les valeurs complexes (6.6). Comme l'intersection des I/K est 0, on voit que $J = Bp$, d'où $K'_S = Bp \otimes A$. Cela prouve (ii). On a alors

$(U(\underline{g})/K)_{S^\sim} = U(\underline{p})_S/K'_S = (B \otimes A)/(Bp \otimes A) = k[q] \otimes A$ ce qui prouve (iii).

La formule (1) s'écrit maintenant $(I/K)_{S^\sim} = k[q](3q+\xi) \otimes A$, d'où (iv).

6.8. Théorème. L'application ν de $\mathfrak{J}_2$ dans $\mathbb{C}$ est bijective.

Cette application est surjective d'après 6.6. Adoptons les notations de 6.7. Soient $I, I' \in \mathfrak{J}_2$ tels que $\nu(I) = \nu(I')$. Alors $(I/K)_{S^\sim} = (I'/K)_{S^\sim}$ d'après 6.7(iv), d'où $I/K = I'/K$ ([5], 3.6.15(ii)), et $I = I'$.

6.9. L'invariant $\nu(I)$ assure donc une paramétrisation parfaite des éléments de $\mathfrak{J}_2$. Voyons maintenant les relations entre cet invariant, et les invariants $\lambda(I)$, $\mu(I)$; ces derniers, bien adaptés pour l'étude de $\mathfrak{J}_1$, le sont beaucoup moins pour l'étude de $\mathfrak{J}_2$:

6.10. Théorème. (i) Soient $I \in \mathfrak{J}_2$ et $\nu = \nu(I)$. Alors $\lambda(I) = \nu^2 - \frac{9}{4}$ et $\mu(I) = 2\nu(\nu^2 - \frac{9}{4})$.

(ii) Soit $(\lambda, \mu) \in \mathbb{C}^2$.

a) Si $\mu^2 \neq 4\lambda^3 + 9\lambda^2$, il n'existe aucun $I \in \mathfrak{J}_2 \cup \mathfrak{J}_3$ tel que $\lambda = \lambda(I)$, $\mu = \mu(I)$.

b) Si $\mu^2 = 4\lambda^3 + 9\lambda^2$ et $(\lambda, \mu) \neq (0,0)$, il existe un $I \in \mathfrak{J}_2$ et un seul tel que $\lambda = \lambda(I)$, $\mu = \mu(I)$. On a $\lambda \neq 0$ et $\nu(I) = \frac{1}{2}\mu\lambda^{-1}$.

c) Il existe deux éléments I de $\mathfrak{J}_2$ tels que $\lambda(I) = \mu(I) = 0$. Les valeurs de ν pour ces deux idéaux sont $3/2$ et $-3/2$.

D'après 6.8 et 6.6, il suffit de vérifier (i) pour les idéaux du lemme 6.6; dans ce cas, il n'y a qu'à faire usage des formules de ce lemme. L'assertion (ii) résulte de (i) par des calculs d'élimination faciles.

6.11. L'espace des idéaux primitifs complètement premiers de $U(\underline{g})$ peut donc être représenté par la figure 2. (Voir figure 2 en annexe).

On voit que les espaces $\underline{g}/G$ et $\mathfrak{J}$ ne sont pas homéomorphes.

6.12. Si $I \in \mathfrak{J}_1$, $U(\underline{g})/I$ se plonge dans l'algèbre de Weyl A_3 avec même corps des fractions [3]. Si $I \in \mathfrak{J}_2$, la démonstration 6.7 prouve que $\mathrm{Fract}\,(U(\underline{g})/I) = \mathrm{Fract}(A_2)$. D'après [3], $U(\underline{g})/I$ se plonge dans A_2.

7. Induction et idéaux primitifs complètement premiers.

7.1. Théorème. (i). Pour tout $f \in \underline{g}^*$, il existe une polarisation de $\underline{g}$ en f.

(ii) <u>Soient $f \in \underline{g}^*$, $\underline{p}_1$ et $\underline{p}_2$ des polarisations de $\underline{g}$ en f. Alors $\mathrm{ind}^\sim(f|\underline{p}_1,\underline{g})$ et $\mathrm{ind}^\sim(f|\underline{p}_2,\underline{g})$ ont un même noyau dans $U(\underline{g})$. Notons $I(f)$ ce noyau.</u>

(iii) <u>L'application I définit une bijection $\bar{I}$ de $\underline{g}^*/G$ sur $\mathfrak{J}$. Cette bijection n'est pas continue.</u>

(iv) <u>On a $\bar{I}(\mathcal{O}_1) = \mathfrak{J}_1$, $\bar{I}(\mathcal{O}_2) = \mathfrak{J}_2$, $\bar{I}(\mathcal{O}_3) = \mathfrak{J}_3$.</u>

Les assertions du théorème sont bien connues pour $\mathcal{O}_1$ et évidentes pour $\mathcal{O}_3$ (mise à part la non continuité). (L'assertion(i) est aussi bien connue pour tout f; cf. par exemple [9]).

Soit $f \in \mathcal{O}_2$ et prouvons (i),(ii). Par automorphisme intérieur, on se ramène aux deux cas suivants :

a) $f(X_\alpha) = f(X_\beta) = f(X_\gamma) = f(X_{-\alpha}) = f(X_{-\beta}) = f(X_{-\gamma}) = f(H_\alpha) = 0, f(H_\gamma) = \xi \neq 0.$

Le noyau de la forme bilinéaire alternée associée à f est alors $\mathbb{C}X_\alpha + \mathbb{C}X_{-\alpha} + \mathbb{C}H_\alpha + \mathbb{C}H_\gamma$. Il y a donc deux polarisations, $\underline{p}$ et $\underline{q} = \underline{h} + \mathbb{C}X_{-\alpha} + \mathbb{C}X_{-\beta} + \mathbb{C}X_{-\gamma} + \mathbb{C}X_\alpha$. Alors $\mathrm{ind}^\sim(f|\underline{p},\underline{g})$ est la représentation étudiée en 6.6, avec ξ remplacé par $\xi - \frac{1}{2}(\beta+\gamma)(H_\gamma) = \xi - \frac{3}{2}$ puisqu'il s'agit d'induction tordue. Le noyau de cette représentation est donc l'élément I de $\mathfrak{J}_2$ tel que $\nu(I) = \xi$. Des calculs analogues à ceux de 6.6 prouvent que le noyau de $\mathrm{ind}^\sim(f|\underline{q},\underline{g})$ est un élément I' de $\mathfrak{J}_2$ tel que $\nu(I') = \xi$, d'où I = I'.

b) $f(X_\alpha) = f(X_\gamma) = f(X_{-\alpha}) = f(X_{-\beta}) = f(X_{-\gamma}) = f(H_\alpha) = f(H_\gamma) = 0$, $f(X_{-\beta}) = 1$.

Le noyau de la forme bilinéaire alternée associée à f est alors $\mathbb{C}(H_\alpha - H_\gamma) + \mathbb{C}X_\alpha + \mathbb{C}X_\beta + \mathbb{C}X_\gamma$. L'unique sous-algèbre de Borel contenant ce noyau est $\underline{h} + \mathbb{C}X_\alpha + \mathbb{C}X_\beta + \mathbb{C}X_\gamma$, d'où deux polarisations de $\underline{g}$ en f, à savoir $\underline{p}$ et $\underline{r} = \underline{h} + \mathbb{C}X_\alpha + \mathbb{C}X_\beta + \mathbb{C}X_\gamma + \mathbb{C}X_{-\gamma}$. Alors $\mathrm{ind}^\sim(f|\underline{p},\underline{g})$ est la représentation étudiée en 6.6, avec ξ remplacé par $-\frac{3}{2}$.

Le noyau de cette représentation est donc l'élément I de $\mathfrak{J}_2$ tel que $\nu(I) = 0$. Des calculs analogues à ceux de 6.6 prouvent que le noyau de $\text{ind}^{\sim}(f|\underline{r},\underline{g})$ est ce même élément de $\mathfrak{J}_2$.

Cela achève la démonstration de (i) et (ii). Du même coup, et compte tenu de 6.8, on a prouvé que $\bar{I}|\mathcal{O}_2$ est une bijection de $\mathcal{O}_2$ sur $\mathfrak{J}_2$.

La comparaison de 2.2 et 6.11 prouve qu'il existe des points de $\mathcal{O}_1$ non fermés dans $\underline{g}/G$ dont l'image par $\bar{I}$ est un point fermé de $\mathfrak{J}$. Donc $\bar{I}$ n'est pas continu.

ANNEXE :

Figure 1

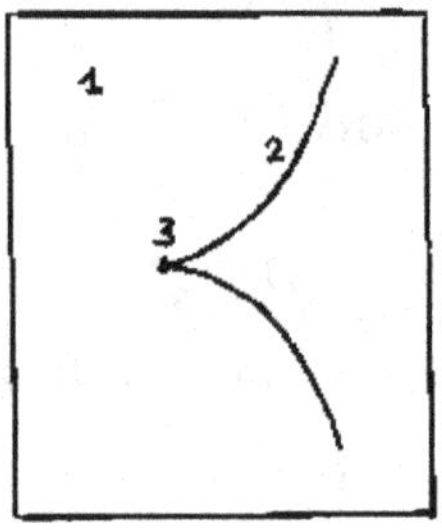

Figure 2

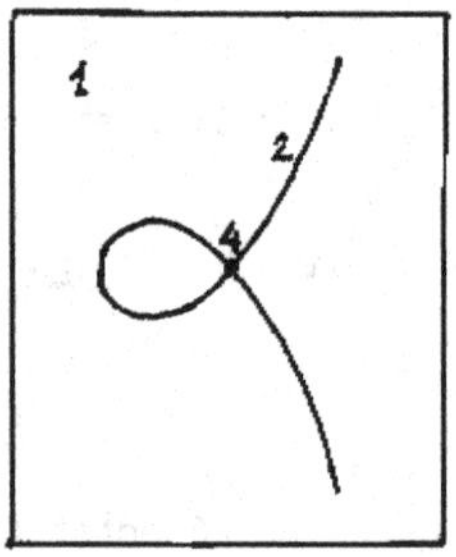

BIBLIOGRAPHIE

[1] W. Borho, P.Gabriel et R. Rentschler, Primideale in Einhüllenden auflösbaren Lie-Algebren, Lecture Notes in Math., 357(1973).

[2] N. Conze, Espace des idéaux primitifs de l'algèbre enveloppante d'une algèbre de Lie Nilpotente, à paraître au J. of algebra.

[3] N. Conze, Algèbres d'opérateurs différentiels et quotients des algèbres enveloppantes, à paraître au Bull. Soc. Math.France.

[4] N. Conze et M. Duflo, Sur l'algèbre enveloppante d'une algèbre de Lie résoluble, Bull.Sci.Math., 94(1970), p.201-208.

[5] J. Dixmier, Algèbres enveloppantes, Cahiers Scient., Paris,Gauthier-Villars, (1974).

[6] J. Dixmier, Idéaux primitifs dans l'algèbre enveloppante d'une algèbre de Lie semi-simple complexe, C.R. Acad. Sci.,272 (1971), p.1628-1630.

[7] N. Jacobson, Structure of rings, Amer. Math. Soc. Coll. Publ.,37,2nd ed.,1964.

[8] B. Kostant, Lie group representations on polynomial rings, Amer. J. Math., 85 (1963), p. 327 - 404.

[9] H. Ozeki et M. Wakimoto, On polarizations of certain homogeneous spaces, Hiroshima Math. J.,2(1972), p.445-482.

[10] R. Rentschler, L'injectivité de l'application de Dixmier pour les algèbres de Lie résolubles, Inv. Math., 23(1974), p.49-71.

Université PARIS VI
Analyse Probabilités et Applications
4 Place Jussieu
75230 PARIS CEDEX 05

SEMIGROUPS OF COMPLEX MEASURES ON A LOCALLY COMPACT GROUP

Michel DUFLO

Here are the main points of a paper which will appear with all details elsewhere. In the first part, I show that there is a bijection between the set of vaguely continuous semi groups $\{\mu_t\}_{t \geq 0}$ of complex Radon measures of total variation ≤ 1 on a locally compact group G, and the set of dissipative distributions T, given by the formula $\langle T, f \rangle = \lim_{t\to 0} t^{-1} \langle \mu_t - \mu_0, f \rangle$ for $f \in D(G)$. This extends results of Hunt, Faraut, Hazod. The method employed gives a new proof for these particular cases.

In the third part, we consider a strongly continuous representation π of G in a banach space H. In some cases, for instance if π is uniformly bounded, or if T has a compact support, it is possible to define a strongly continuous semigroup of bounded operators $\{\pi(\mu_t)\}_{t \geq 0}$ in H. I show that $\pi(T)$ is the infinitesimal generator of this semigroup, for any reasonable definition of $\pi(T)$. This extends and unifies results of Hunt, Nelson-Stinespring, Faraut, Phillips.

I - Semi groups of measures and dissipative distributions.

1. Notations and definitions. We fix a locally compact group G. We use Bruhat's notations, so that $E(G)$ is the set of regular functions on G, $D(G)$ the set of regular functions with compact support. We note $C_o(G)$ the space of continuous functions which vanish at infinity. If μ is a (Radon) measure on G, we note $|\mu|$ its absolute value and $\|\mu\|$ its total variation. We fix an idempotent measure χ on G such that $\|\chi\| = 1$. The support of χ is a compact subgroup K of G, and χ is equal to the Haar measure of K multiplied by an unitary character of K. If u is a distribution, put $d\check{u}(x) = du(x^{-1})$. We note $D(G,\chi)$ the set of $f \in D(G)$ such that $f * \check{\chi} = f$. In the same way, we definie $C_o(G,\chi)$, etc... . We fix a right Haar measure on G, and use it to identify locally

summable functions and measures on G .

We note $M(G,\chi)$ the set of vaguely continuous semigroups $\{\mu_t\}_{t \geq 0}$ of complex measures on G such that $\|\mu_t\| \leq 1$ for all $t \geq 0$, and $\mu_o = \chi$. (Every vaguely continuous semigroup $\{\mu_t\}_{t > 0}$ of complex measures such that $\|\mu_t\| \leq 1$ can uniquely be extended to a semi group in some $M(G,\chi)$).

We call a distribution T χ-dissipative if $\chi * T * \chi = T$, and if $\operatorname{Re} T(f) \leq 0$ for all $f \in D(G,\chi)$ such that $f(1) = \sup_{x \in G} |f(x)|$.

2. Statement of theorem 1.

<u>Theorem</u> 1 - <u>Let</u> $\{\mu_t\} \in M(G,\chi)$. <u>There is a</u> χ-<u>dissipative distribution</u> T <u>such that</u>

(1) $\lim t^{-1} \langle \mu_t - \chi, f \rangle = \langle T, f \rangle$ <u>for all</u> $f \in D(G)$.

<u>The formula</u> (1) <u>defines a bijection between</u> $M(G,\chi)$ <u>and the set of</u> χ- <u>dissipative distributions</u>.

3. The proof of theorem 1 uses the following lemmas. The first one is easy. The proof of the second one use structure theory and Taylor formula.

<u>Lemma</u> 1 - <u>Let</u> T <u>be a</u> χ-<u>dissipative distribution and</u> V <u>be a closed neighborhood of</u> K . <u>Then the restriction of</u> T <u>to</u> $G-V$ <u>is a bounded measure</u>.

Lemma 1 shows that T can be naturally extended to $C_o(G) \cap E(G)$.

<u>Lemma</u> 2 - <u>Let</u> V <u>be a compact neighbo rhood of</u> K <u>and</u> H <u>be a subspace of</u> $C_o(G,\chi) \cap E(G)$, <u>dense for the uniform convergence on</u> G , <u>and the uniform convergence of derivatives on</u> K . <u>Let</u> U <u>be a linear form on</u> H , χ-<u>dissipative. Then there exists a unique</u> χ-<u>dissipative distribution</u> T <u>such that</u> $T(f) = U(f)$ <u>for all</u> $f \in H$.

4. Proof of theorem 1. Let $\{\mu_t\}_{t \geq 0} \in M(G,\chi)$. For $f \in C_o(G,\chi)$, we put $P_t(f) = f * \check{\mu}_t$.

Then $\{P_t\}_{t \geq 0}$ is a strongly continuous semigroup of contractions on $C_o(G,\chi)$.
From lemma 2 it follows that there exists a χ-dissipative distribution T such

that $\alpha * A(f) = \alpha * f * \check{T}$ for all $\alpha \in D(G)$, $f \in \operatorname{dom} A$, where A is the infinitesimal generator of $\{P_t\}$. Let B be the operator $f \longrightarrow f * \check{T}$ on $C_o(G, \chi)$, whose domain is the set of $f \in C_o(G, \chi)$ such that the distribution $f * \check{T}$ is still in $C_o(G, \chi)$. Then B is a dissipative operator which extends A . Since A (as an infinitesimal generator of a contraction semi-group) is a maximal dissipative operator, we get A = B . This proves (1) .

Conversely, let T be a χ-dissipative distribution. Define B as above. A simple application of Hahn-Banach theorem shows that $(B-\lambda)(D(G, \chi))$ is dense in $C_o(G, \chi)$ for every $\lambda > 0$. It follows that B is the closure of its restriction to $D(G, \chi)$, and that B generates a contraction semigroup in $C_o(G, \chi)$. This proves the second part of theorem 1.

5. The proof above gives also the following results.

<u>Proposition</u> 1 - <u>Let</u> $\{\mu_t\}$ <u>and</u> T <u>be as in theorem 1</u>. <u>For</u> $f \in C_o(G, \chi)$ <u>put</u> $P_t(f) = f * \check{\mu}_t$, <u>and let</u> A <u>be the infinitesimal generator of</u> $\{P_t\}$. <u>Then the domain of</u> A <u>is the set of</u> $f \in C_o(G, \chi)$ <u>such that</u> $f * \check{T} \in C_o(G, \chi)$. <u>If</u> $f \in \operatorname{dom} A$, <u>then</u> $A(f) = f * \check{T}$. <u>The operator</u> A <u>is the closure of its restriction to</u> $D(G, \chi)$.

Proposition 1 will be generalized below (theorem 2).

<u>Proposition</u> 2 - <u>Let</u> A <u>be a densely defined left-invariant dissipative operator in</u> $C_o(G, \chi)$. <u>There exists a left-invariant semigroup of contractions in</u> $C_o(G, \chi)$ <u>whose infinitesimal generator extends</u> A .

6. Support of $\{\mu_t\}$.

<u>Proposition</u> 3 - <u>Let</u> C <u>be a closed sub-semigroup of</u> G <u>which contains</u> K . <u>Let</u> $\{\mu_t\}$ <u>and</u> T <u>be as in theorem 1</u>. <u>Then the support of</u> $\{\mu_t\}$ <u>is contained in</u> C <u>if and only if</u> $\operatorname{Re} T(f) \leqslant 0$ <u>for every</u> $f \in D(G, \chi)$ <u>such that</u> $f(1) = \sup_{x \in C} |f(x)|$.

<u>Proof</u>. Suppose the second condition is verified. Put $B(f|C) = f * \check{T}|C$ for $f \in D(G, \chi)$, where $.|C$ means the restriction to C . Then B is a densely defined dissipative operator in $C_o(C, \chi)$, and its closure generates a C - left-

invariant contraction semi-group in $C_o(G,\chi)$. This corresponds to a semigroup of measures on C which is easily identified with $\{\mu_t\}$.

The case where C is a cone in $G = \underline{R}^n$ is due to Faraut, who uses Laplace transform.

7. Markovian semigroups and generalized Laplacians.

Suppose χ is the Haar measure of K .

We note $P(G,\chi)$ the subset of $M(G,\chi)$ which consists of semigroups of positive measures A distribution T is a χ- generalized laplacian on G if $\chi * T * \chi = T$, and if $T(f) < 0$ for any real $f \in D(G,\chi)$ such that $f(1) = \sup_{x \in C} f(x) \geqslant 0$. With the notations of theorem 1, $\{\mu_t\} \in P(G,\chi)$ if and only if T is a χ- generalized laplacian.

We do not suppose any more that χ is positive. We note $\underline{T}$ the circle group. If f is a function on G , we note zf the function on $G \times \underline{T}$ defined by $zf(g,t) = t\, f(g)$. There exists a unique compact subgroup K' of $G \times \underline{T}$ such that its Haar maesure χ' verfies $\chi'(zf) = \chi(f)$ for $f \in D(G)$.

Let $\{\mu_t\}$ and T be as in theorem 1. It has been proved by Roth that there exists $\{\gamma_t\} \in P(G \times \underline{T}, \chi')$ such that $\gamma_t(zf) = \mu_t(f)$ for every $t \geqslant 0$ and $f \in D(G)$. Then, if U is the χ'-generalized laplacian on $G \times \underline{T}$ associated to $\{\gamma_t\}$, we also have $U(zf) = T(f)$ for $f \in D(G)$. (For $G = \underline{R}^n$, this is due to Faraut). This is usefull, because the structure of generalized laplacians is well understood. For instance, we get the following proposition.

<u>Proposition</u> 4 - <u>Let</u> T <u>be a</u> χ-<u>dissipative distribution on</u> G <u>and</u> V <u>a neighborhood of</u> K . <u>There exists a</u> χ'-<u>generalized laplacian</u> U <u>on</u> $G \times \underline{T}$ <u>with support contained in</u> V <u>and a bounded measure</u> T^2 <u>on</u> G <u>such that</u> $T(f) = U(zf) + T^2(f)$ <u>for all</u> $f \in D(G)$.

Put $T^1(f) = U(zf)$, and let $\{\mu_t\}$ and $\{\mu_t^1\} \in M(G,\chi)$ correspond to T and T^1 . Then μ_t can be obtained from μ_t^1 by mean of the pertubation series

(2) $$\mu_t = \mu_t^1 + \int_o^t \mu_{t-u}^1 * T^2 * \mu_u^1 \, du + \ldots..$$

This reduces number of questions relative to $\{\mu_t\}$ to analoguous questions relative to semigroups in $P(G \times \underline{T}, \chi')$ with compactly supported generator.

Let C be a closed sub-semigroup which contains the support of $\{\mu_t\}$. Then we can suppose that $\{\gamma_t\}$ is supported by $C \times \underline{T}$. We can suppose then that T^1 and T^2 have supports contained in C.

II - Generalized laplacians with compact support.

1. We suppose that χ is the Haar measure of K. We fix $\{\mu_t\} \in P(G, \chi)$ and note T the corresponding χ- generalized laplacian. We suppose that the support of T is contained in an open relatively compact neighborhood V of K. We suppose that G is separable.

Proposition 5 - *There exists* $c > 0$ *such that for all* $t \in [0,1]$ *and* $n \in \underline{N}$, $n \geqslant 1$, *we have* $\mu_t(G - V^{4n}) \leqslant c^n t^n n^{-n}$.

The proof uses probabilistic methods. Associated to the semi group, there is on G/K a strong Markov process with trajectories continuous on the right, with limit on the left, and bounded jumps.

2. Let C be a closed sub-semigroup of G which supports $\{\mu_t\}$. Let φ be a weight on C, i.e. a measurable locally bounded submultiplicative function on C With values in $(0, \infty)$. Let $f \in E(G)$ verify $|f(x)| \leqslant \varphi(x)$ for all $x \in C$. Then we have :

$$\lim_{t \to 0} t^{-1} \langle \mu_t - \chi, f \rangle = \langle T, f \rangle . \tag{3}$$

In fact, the left-hand side of (3) is defined by proposition 5, and (3) is obtained by writing $f = f_1 + f_2$, with $f_1 \in D(G)$, and f_2 mull in V^8.

We put $\omega = \lim_{t \to \infty} t^{-1} \log (\mu_t(\varphi))$.

Proposition 7 - *Let* $f \in E(G)$ *verify the following conditions.* 1 . *There exists a locally bounded positive function* c *on* G *such that* $|f(xy)| \leqslant c(x)\varphi(y)$ *for* $x \in G$, $y \in C$. 2 . *For all* $t > 0$, $f * \check{\mu}_t$ *(which is defined by* 1) *belongs to* $E(G)$. 3 . *There exists* $\lambda > \omega$ *such that* $f * \check{T}(x) = \lambda f(x)$ *for all* $x \in C$.

<u>Then</u> $f(x) = 0$ <u>for all</u> $x \in C$.

<u>Proof</u>. From (3) , we get that $e^{-\lambda t} \; f * \check{\mu}_t \,(x) = f(x)$ for all $x \in C$, $t \geqslant 0$. As t tends to ∞ , the left hand side tends to 0 , and thus $f(x) = 0$.

In the particular case where $f|C \in C_o(C)$, proposition 7 is a simple consequence of the maximum principle (proposition 3). When $C = G$ (or more generally if the interior C of C is dense in C) condition 2 in proposition 7 is not necessary.

III - <u>Representations of semigroups</u>.

1. We do not suppose any more that χ is positive. We fix $\{\mu_t\} \in M(G, \chi)$ and note T the corresponding χ-dissipative distribution. We fix a closed sub-semigroup C of G which supports $\{\mu_t\}$ and such that the interior $\mathring{C}$ of C is dense in C . We fix a strongly continuous representation π of C in a banach space H . The function $\varphi(x) = \|\pi(x)\|$ is a weight on C . Let V be a compact neighborhood of K . We suppose that we have :

$$\int_{C-V} \varphi(x) \;\; d|T|(x) < \infty \; . \tag{4}$$

(Recall that T is a bounded measure outside V). Condition (4) is verified for instance if π is uniformly bounded, or T with compact support.

Formula (2) and proposition 5 shows that $|\mu_t|(\varphi) < \infty$ for all $t \geqslant 0$. We can then define

$$\pi(\mu_t) = \int_G \varphi(x) \; d\mu_t(x) \; .$$

Then $\{\pi(\mu_t)\}_{t \geqslant 0}$ is a strongly continuous semigroup of bounded operators in H . We shall note A its generator : $A = \underset{t \to 0}{\text{s-lim}}\; t^{-1}(\pi(\mu_t) - \pi(\chi))$.

2. Image of distributions.

Let u be a distribution on G whose support is compact and contained in C . We note $D(C)$ the set of $\alpha \in D(G)$ whose support is contained in C . The

Garding's subspace H' is the subspace of H generated by the $\pi(\alpha)h$ with $\alpha \in D(C)$ and $h \in H$. Since 1 is adherent to $\mathring{C}$, H' is dense in H. We define an operator $\pi_2(u)$ in H in the following way. The domain of $\pi_2(u)$ is the set of $h \in H$ such that there exists $k \in H$ such that

$$\pi(\alpha * u)\,h = \pi(\alpha)\,k$$

for all $\alpha \in D(C)$. Then k is uniquely determined and we set $\pi_2(u)\,h = k$. The space H' is contained in the domain of $\pi_2(u)$: in fact

$$\pi_2(u)\,\pi(\alpha)\,h = \pi(u * \alpha)\,h$$

for $h \in H$ and $\alpha \in D(C)$.

The operator $\pi_2(u)$ is closed and densely defined. We note $\pi_1(u)$ the closure of its restriction to H'. If u is central, it is easy to see that $\pi_1(u) = \pi_2(u)$, but this is not the case in general.

3. Statement of theorem 2.

We write $T = T^1 + T^2$ as in proposition 4. By (4), $\pi(T^2)$ is a bounded operator in H. We define $\pi_1(T) = \pi_1(T^1) + \pi(T^2)$ and $\pi_2(T) = \pi_2(T^1) + \pi(T^2)$. Then $\pi_1(T)$ and $\pi_2(T)$ are densely defined closed operators in H, and $\pi_2(T)$ extends $\pi_1(T)$.

Theorem 2 - We keep the notations and hypothesis of III.1. Then $A = \pi_1(T) = \pi_2(T)$.

Examples. 1) Theorem 2, applied to the right regular representation of G in $C_o(G)$, reduces exactly to proposition 1.

2) The inclusion $\pi_1(T) \subset A$, applied to the right regular representation of G in the space of right uniformly continuous functions on G is due to Hunt (when $\{\mu_t\} \in P(G,\chi)$).

3) When π is uniformly bounded, G a Lie group, $C = G$, and the operator $f \longrightarrow f * \check{T}$ $(f \in D(G))$ a second order elliptic differential operator, the equality $\pi_1(T) = A$ is due to Nelson and Stinespring.

4) Suppose $C = G$, and suppose that π is a unitary representation in a Hilbert space. Then $\pi_1(\check{u}) = \pi_2(u)^*$ for every distribution u with compact

support. Thus $\pi_1(T)^* = \pi_1(\check{T})$. For other examples of this phenomena, see Nelson and Stinespring.

5) The case $G = \underline{R}$, $C = [0, \infty)$ has been investigated by Phillips and Faraut.

4. Proof of theorem 2.

The inclusions $\pi_1(T) \subset A \subset \pi_2(T)$ are easy to prove - see for example Faraut p. 283. We use perturbation methods and proposition 4 to show that it is enough to prove theorem 2 when T is a compactly supported generalized laplacian. We suppose that this is the case. Define ω as before proposition 7. To prove that $A = \pi_1(T)$, it is enough to prove that for some $\lambda > \omega$, $(\pi_1(T) - \lambda) H'$ is a dense subspace of H . Let ξ be a continuous linear form null on this space. Let $h \in H$, put $f'(x) = \langle \xi, \pi(x) h \rangle$ $(x \in C)$. Let $\beta \in D(C)$ and put $f = f' * \check{\alpha}$. Let $\alpha \in D(C)$. We have $\langle \xi, (\pi_1(T) - \lambda) \pi(\alpha) \pi(\beta) h \rangle = 0$, and thus $\langle f * \check{\alpha}, T - \lambda \rangle = 0$. Put $\tilde{f}(x) = f(x^{-1})$ $(x \in G)$. Then, $\langle \tilde{f} * (T - \lambda), \check{\alpha} \rangle = 0$. Since α is arbitrary in $D(C)$, this proves that $\tilde{f} * T(x) = \lambda f(x)$ for all $x \in C^{-1}$. Proposition 7, applied to $\check{T}$ and C^{-1} , shows that $f = 0$. Since β and h are arbitra ry, we obtain $\xi = 0$. Thus $A = \pi_1(T)$.

To prove that $A = \pi_2(T)$, it is enough to show that for some $\lambda > \omega$ the operator $\pi_2(T) - \lambda$ is injective. Suppose that h is an element of dom $\pi_2(T)$ such that $(\pi_2(T) - \lambda) h = 0$. Let ξ be a linear continuous form on H, $\alpha, \beta \in D(C)$. Then $\langle \xi, \pi(\beta * \alpha * (T - \lambda)) h \rangle = 0$.

Define f' as above, and put $g(x) = \int_C f'(xy) \beta(y)\, dy$.

We get $\langle g * (\check{T} - \lambda), \alpha \rangle = 0$. As above we conclude that $h = 0$.

References :

F. BRUHAT - Distributions sur un groupe localement compact et applications à l'étude des représentations des groupes p-adiques. Bull. Soc. Math. Fr. 89 (1961) 43-75.

J. FARAUT - Semi-groupes de mesures complexes et calcul symbolique sur les générateurs infinitésimaux de semi-groupes d'opérateurs. Ann. Inst. Fourier. 20 (1970) 235-301.

W. HAZOD - Uber die Lévy-Hincin-Formel auf lokalkompackten Gruppen. Z. W ahrscheinlich keitstheorie verw. Geb. 25 (1973) 301-322.

G. A. HUNT - Semigroups of measures on Lie groups. Trans. Amer. Math. Soc. 81 (1956) 264-293.

E. NELSON and W. F. STINESPRING - Representation of elliptic operators in an enveloping algebra. Amer. J. Math. 81 (1959) 547-560.

R.S. PHILLIPS - On the generation of semigroups of linear operators. Pacific J. Math. 2 (1952) 343-369.

J.P. ROTH - Sur les semi-groupes à contraction invariants sur un espace homogène. C.R. Acad. Sc. Paris 277 (1973) 1091-1094.

Université PARIS VII
U. E. R de Mathématiques
2 Place Jussieu
75221 PARIS CEDEX 05

SPHERICAL FUNCTIONS AND DISCRETE SERIES

Mogens Flensted-Jensen

Introduction. If G is a "nice" locally compact group the Plancherel formula

$$(1) \qquad \|f\|_2^2 = \int_{G^\wedge} \mathrm{Tr}(\pi(f)\pi(f)^*)\,d\mu(\pi)$$

for $f \in L^1(G) \cap L^2(G)$ is abstractly known, (see e.g. Dixmier [2], theorem 18.8.2). If H is a compact subgroup of G and $f \in C_c(G/H)$, then it is easily seen, that the function $\pi \to \mathrm{Tr}(\pi(f)\pi(f)^*)$ has support in the set

$$G^\wedge(H) = \{\pi \in G^\wedge \mid \pi(H) \text{ has a non-zero fixed vector}\} .$$

The restriction of the Plancherel measure μ to $G^\wedge(H)$ is the Plancherel measure for G/H.

If the convolution algebra

$$C^\natural(G,H) = \{f \in C_c(G) \mid f(h\,xh_1) = f(x) \text{ for all } h, h_1 \in H\}$$

is commutative, then the set $G^\wedge(H)$ is known to be in one-one correspondance with the set of positive definite spherical functions φ on G w.r.t. H, satisfying $\varphi(e) = 1$. (Godement [5]).

In this paper we are especially interested in studying, when a representation corresponding to a spherical function is in the relative discrete series of the group. For some of the simply connected, non-compact, semi-simple Lie groups, with the right choice of a compact subgroup, we shall in this way find a nice subset of the discrete series. The following lemma is useful for this:

Lemma 1. Let $C^{\natural}(G,H)$ be commutative, and let φ be a spherical function. Then φ is positive definite and corresponds to a representation of the relative discrete series if and only if φ satisfies:

(a) φ restricted to the center, $Z(G)$, is a unitary character δ of $Z(G)$.

(b) $|\varphi| \in L^2(G/Z(G))$.

Proof: (a) and (b) are clearly necessary So assume (a) and (b) satisfied. Let τ be the induced representation $\operatorname{ind}_{Z(G)\uparrow G}(\delta)$. Clearly $\overline{\varphi}$ is in the Hilbert space for τ and

$$(\tau(x)\overline{\varphi}, \overline{\varphi}) = \overline{\varphi(x^{-1})}\,\|\overline{\varphi}\|^2 \quad .$$

This shows that φ is positive definite and the sufficiency follows.

Q.e.d.

For G semi-simple, non-compact, with finite center, the approach by means of spherical functions has been used very successfully by Harish-Chandra and others, with H taken as the maximal compact subgroup K of G. Let us briefly recall how Harish-Chandra describes the Plancherel measure for G/K (see Harish-Chandra [6] and [7]): Let $\mathfrak{g} = \mathfrak{k} + \mathfrak{a} + \mathfrak{n}$ and $G = KAN$ be the Iwasawa decomposition of the Lie algebra $\mathfrak{g}$ of G, and of G. For $x \in G$ let $H(x)$ be defined as the unique element in $\mathfrak{a}$ such that $x \in K\exp(H(x))N$. The spherical functions on G w.r.t. K

are parametrized by means of the complex dual $\mathfrak{a}_{\mathbb{C}}^*$ of $\mathfrak{a}$, (modulo the Weyl group W):

$$\varphi_\lambda(x) = \int_K e^{(i\lambda-\rho)(H(xk))}dk, \quad \lambda \in \mathfrak{a}_{\mathbb{C}}^*$$

where ρ is half the sum of the restricted roots.

For "most" λ the spherical function φ_λ has an expansion:

$$\varphi_\lambda = \sum_{w\in W} c(w\lambda)\Phi_{w\lambda}, \tag{2}$$

where c is Harish-Chandras "c-function", which is explicitly known, and $\Phi_{w\lambda}$ are defined on A^+, with known asymptotic behavior. The Plancherel measure for G/K is then $|c(\lambda)|^{-2}d\lambda$ over $\mathfrak{a}_{\mathbb{R}}^*$. In this case there are no discrete series corresponding to spherical functions. In the following we shall indicate an extension of this approach, which comes about by choosing H to be a smaller compact subgroup of G. It only gives something new for the case, when the universal covering group $G^\sim$ of G has infinite center, or equivalently, when the subgroup K has non-discrete center. If we also assume that G is simple, this happens precisely when G/K is a Hermitian symmetric space. We must now introduce some more notations, and make our assumptions on G precise:

Section 1. Let $\mathfrak{g}$ be a simple, non-compact Lie algebra, with Cartan-decomposition $\mathfrak{g} = \mathfrak{k} + \mathfrak{p}$ and Iwasawa-decomposition $\mathfrak{g} = \mathfrak{k} + \mathfrak{a} + \mathfrak{n}$. $\mathfrak{k}$ is the direct sum of a compact, semi-simple Lie algebra $\mathfrak{k}_0$ and the center $\mathfrak{k}_1$ of $\mathfrak{k}$, where $\dim \mathfrak{k}_1 = 1$. Let $\mathfrak{m}$ be the centralizer of $\mathfrak{a}$ in $\mathfrak{k}$. Choose a Cartan-subalgebra

$\mathfrak{h}$ of $\mathfrak{k}$. Let $\mathfrak{g}_{\mathbb{C}}$, $\mathfrak{k}_{\mathbb{C}}$ and $\mathfrak{h}_{\mathbb{C}}$ be the complexifications. Choose a Weyl basis $\{E_\alpha\}_{\alpha\in\Delta}$ of $\mathfrak{g}_{\mathbb{C}}$ modulo $\mathfrak{h}_{\mathbb{C}}$ with respect to the compact real form $\mathfrak{u} = \mathfrak{k} + i\mathfrak{p}$ (see Helgason [8], IX, §2) . We can assume (Helgason [8], VIII, cor. 7.6) , that

$$\mathfrak{a} = \sum_{i=1}^{r} \mathbb{R}(E_{\beta_i} - E_{-\beta_i}) ,$$

where $r = \dim$ and $\beta_1,\cdots,\beta_r$ is a set of strongly orthogonal non-compact roots.

Let G be a connected Lie group with Lie algebra $\mathfrak{g}$. Assume, for simplicity, that G is simply connected. Let K, K_0, K_1, A and N be the analytic subgroups of G corresponding to $\mathfrak{k}$, $\mathfrak{k}_0$, $\mathfrak{k}_1$, $\mathfrak{a}$ and $\mathfrak{n}$. Since G is simply connected, K is a direct product of K_0 and K_1 . K is not compact since K_1 is isomorphic to $(\mathbb{R},+)$, and K_0 is a maximal compact subgroup of G . It is then very natural to take $H = K_0$ and to ask whether the spherical function approach works. By the remarks in the introduction we shall look at $G^\wedge(K_0)$ and $C^\natural(G,K_0)$.

Let $K_1^\wedge$ denote the set $\{\delta \in K^\wedge \mid \dim \delta = 1\}$. Since every $\delta \in K_1^\wedge$ is trivial on K_0 it follows that $K_1^\wedge$ is just the character group of K_1 . If $\pi \in G^\wedge$ and $\delta \in K^\wedge$ we mean by $\delta \in \pi_{|K}$ that δ is a direct summand in the restriction of π to K .

<u>Proposition 2</u>. $G^\wedge(K_0)$ equals the set of $\pi \in G^\wedge$ such that $\delta \in \pi_{|K}$ for some $\delta \in K_1^\wedge$.

The proof is obvious. The following proposition tells us when $C^\natural(G,K_0)$ is commutative.

Proposition 3. Define

$$C^b(G) = \{f \in C^\natural(G,K_0) \mid f(k\,xk^{-1}) = f(x) \text{ for all } x \in G\ ,\ k \in K_1\}\ .$$

$C^b(G)$ is a commutative subalgebra of $C^\natural(G,K_0)$ moreover the following four statements are equivalent :

(i) $C^\natural(G,K_0)$ is commutative.

(ii) $C^\natural(G,K_0) = C^b(G)$

(iii) $\mathfrak{k}_0 + \mathfrak{m} = \mathfrak{k}$

(iv) G/K is not tube type.

Proof: Define an involutive automorphism γ on $\mathfrak{g}_{\mathbb{C}}$ by γ being minus the identity on $\mathfrak{h}_{\mathbb{C}}$ and $\gamma(E_\alpha) = E_{-\alpha}$ for all $\alpha \in \Delta$. Clearly γ leaves $\mathfrak{g}$ and $\mathfrak{k}$ invariant, and also the restriction of γ to $\mathfrak{a} + \mathfrak{k}_1$ equals minus the identity . γ extends to an involution of G . Since it is easily checked that for $f \in C^b(G)$ and $x \in G$ $f(\gamma(x)) = f(x^{-1})$ it follows that $C^b(G)$ is commutative. The proof that (i) - (iv) are equivalent is rather straight forward.

Q.e.d.

Corollary 4. Let G be such that G/K is nontube type. If for $\pi \in G^\wedge$ there is a $\delta \in K_1^\wedge$ such that $\delta \in \pi_{|K}$, then δ is unique and only contained once.

Proof: Since $C^\natural(G,K_0)$ is commutative, it is a standard argument to show, that the dimension of the set of fixvectors for $\pi(K_0)$ is less than or equal to 1 .

Q.e.d.

For the non-tube type cases we have thus seen that the spherical function approach works with $H = K_0$. But before we turn to describe that in more details, we will show that a little trick can include also the tube-type case in the treatment. We extend G by the inner automorphisms of G coming from elements of K_1, in the following sense:

Let L be "another copy" of K_1, define

$$Q = \{(x,x) \mid x \in K_1 \cap Z(G)\}$$

$$G^1 = (G \times L)/Q$$

$$K_1{}^1 = \{(x,x)Q \mid x \in K_1\} \quad \text{and}$$

$$K^1 = K_0 \cdot K_1{}^1$$

Proposition 5. K^1 is compact. The mapping

$$\Psi: gK_0 \to gK^1$$

is a diffeomorphism of G/K_0 onto G^1/K^1 which induces an isomorphism of $C^\natural(G^1, K^1)$ onto $C^b(G)$. In particular is $C^\natural(G^1, K^1)$ commutative.

The proof is straight forward.

By working with spherical functions on G^1 w.r.t. K^1, instead of on G w.r.t. K_0, we can by means of this proposition include the tube-type case in over treatment. The diffeomorphism ψ relates the Plancherel measure for G/K_0 very closely to the Plancherel measure for G^1/K^1. State another way: $(G^1)^\wedge(K^1)$ can be described by means of spherical functions, the restriction from G^1 to G determines a surjection of $(G^1)^\wedge(K^1)$ onto $G^\wedge(K_0)$. So proposition 2 tells us that in this way we can describe all $\pi \in G^\wedge$, which contain a one-dimensional representation δ of K. For the tube-type case

δ is, however, not uniquely determined by π , even the multiplicity of δ must be one. More precisely we have the following result, which is easy to prove.

Proposition 6. Define a "fiber"-space over $G^\wedge(K_0)$ by

$$B(G^\wedge, K_0) = \{(\pi,\delta) \in G^\wedge \times K_1^\wedge \mid \delta \in \pi_{|K}\} .$$

The mapping $(\pi,\delta) \to \pi \otimes \bar{\delta}$ defines a bijection of $B(G^\wedge, K_0)$ onto $(G^1)^\wedge(K^1)$.

Section 2. We are now left with four main problems:

I Find all spherical functions on G^1 w.r.t. K^1 .

II Determine which of the spherical functions are positive definite, in particular, which of them belong to $L^2(G^1/Z(G^1))$

III Find explicitly the Plancherel measure for G^1/K^1 .

IV Determine the fiber over each point of $G^1(K_0)$ in $B(G^\wedge, K_0)$.

In the general case we can solve I and say something about II and III. For the case of rank $G/K = 1$, (i.e. dim $\mathfrak{a} = 1$) we can solve I, III, IV and most of II . But first the general case.

Let ℓ be the Lie algebra of L and let $\mathfrak{a}_1 = \mathfrak{a} + \ell$. For $x \in G^1$ let $H^1(x)$ be the unique element in $\mathfrak{a}_1$ such that $x \in K^1 \exp H^1(x)N$. Extend ρ from $\mathfrak{a}$ to $\mathfrak{a}_1$ by $\rho(\ell) = \{0\}$. Here is the answer to problem I:

Theorem 7. Every spherical function on G^1 w.r.t. K^1 is of the form φ_ν for $\nu \in (\mathfrak{a}_1)_{\mathbb{C}}^*$, where

$$\varphi_\nu(x) = \int_{K^1} e^{(i\nu-\rho)H^1(xk))}dk \quad .$$

Let $\nu_1 = (\mu_1, \lambda_1)$, $\nu_2 = (\mu_2, \lambda_2)$ where $\mu_i \in \mathfrak{a}_{\mathbb{C}}^*$ and $\lambda_i \in \ell_{\mathbb{C}}^*$, then $\varphi_{\nu_1} = \varphi_{\nu_2}$ if and only if $\lambda_1 = \lambda_2$ and $\mu_1 \in W \cdot \mu_2$

The proof is fairly long, but it follows rather closely the proof of Harish-Chandra for the case mentioned in the introduction (see e.g. Helgason [8], X) .

For "most" $\nu = (\mu, \lambda)$ we can find an expansion

$$\varphi_\nu = \sum_{w \in W} c_1(w \cdot \mu \,, \lambda)\Phi_{(w \cdot \mu, \lambda)} \quad ,$$

similar to formula (2) . The new c_1-function can as the c-function in (2) be computed as a product of certain rank-one c_1-functions . It can be considered as a kind of analytic continuation of the c-function since $c_1(\mu,0) = c(\mu)$, where $c(\mu)$ is the c-function corresponding to $G/Z(G)$, $K/Z(G)$. By examining the zeroes and pools of the c_1-function one should be able to find at least "most" of the φ_ν's which belong to $L^2(G^1/Z(G^1))$, since the asymptotic behavior of Φ_ν is known.

For the continuous part of the spectrum the natural conjecture is that the Plancherel-measure should be $|c_1(\nu)|^{-2}\, d\nu$ over $(\mathfrak{a}_1)_{\mathbb{R}}^*$.

Section 3. For the rank 1 case G can, by the classification, only be the universal covering group $SU(n,1)^\sim$ of one of the groups $SU(n,1)$, $n = 1, 2, \cdots$. In this case $K = S(U(n) \times U(1))^\sim$ and $K_0 = SU(n)$. The corresponding Hermitian symmetric space

$SU(n,1)/S(U(n)\times U(1))$ is of tube type only for $n = 1$. In that particular case $K_0 = \{e\}$ and therefore $G^\wedge(K_0) = G^\wedge$, and our method of going to G^1, and using spherical functions there and going back to G, supply us with the full unitary dual of G and the explicit Plancherel measure for the whole universal covering group of $SU(1,1)$ (which is the same as the universal covering groups of $SL(2,\mathbb{R})$.) Notice in passing that the extension procedure on $G = SU(n,1)$, $K_0 = SU(n)$ give $G^1 = U(n,1)$, $K^1 = U(n)$. If we suitably identify $\nu \in (\mathfrak{a}_1)^*_{\mathbb{C}}$ with $(\mu,\lambda) \in \mathbb{C}^2$ and $X \in \mathfrak{a}_1$ with $(t,\theta) \in \mathbb{R}^2$, we find that φ_ν is essentially a hypergeometric function:

$$\varphi_\nu(\exp X) = e^{i\lambda\theta}(\text{ch}t)^\lambda\ {}_2F_1(\tfrac{1}{2}(n+\lambda+i\mu),\ \tfrac{1}{2}(n+\lambda-i\mu);n;-(\text{sh}t)^2)\ .$$

The c_1-function becomes:

$$c_1(\mu,\lambda) = \frac{2^{(n-i\mu)}\ \Gamma(i\mu)\ \Gamma(n)}{\Gamma(\tfrac{1}{2}(n+\lambda+i\mu))\ \Gamma(\tfrac{1}{2}(n-\lambda+i\mu))}\ .$$

The asymtotic behavior of $\Phi_{\mu,\lambda}(t,\theta)$ as $t,\theta\to\infty$ is $e^{i\lambda\theta}\ e^{(i\mu-n)t}$. The only nontrivial Weyl group element takes μ into $-\mu$. Since $\varphi_{\mu,\lambda} = \varphi_{-\mu,\lambda}$ we can assume that $\eta = \text{Im}(\mu) > 0$ or if $\eta = 0$ that $\mu \geqq 0$. It is now easy to see that $\varphi_{\mu,\lambda} \in L^2(G/Z(G))$ if and only if $\lambda \in \mathbb{R}$, $\eta > 0$ and $\mu c_1(-\mu,\lambda) = 0$. This last equation is easily solved, with the following solution:

$$D = \{(\mu,\lambda)\mid -i\mu = \eta > 0\ ,\ \lambda\in\mathbb{R}\ ,\ \eta = -2m-n\pm\lambda \quad \text{for}\quad m = 0,1,\cdots\}$$

The following graph gives a picture of D, and of the set E defined below:

Figure 1

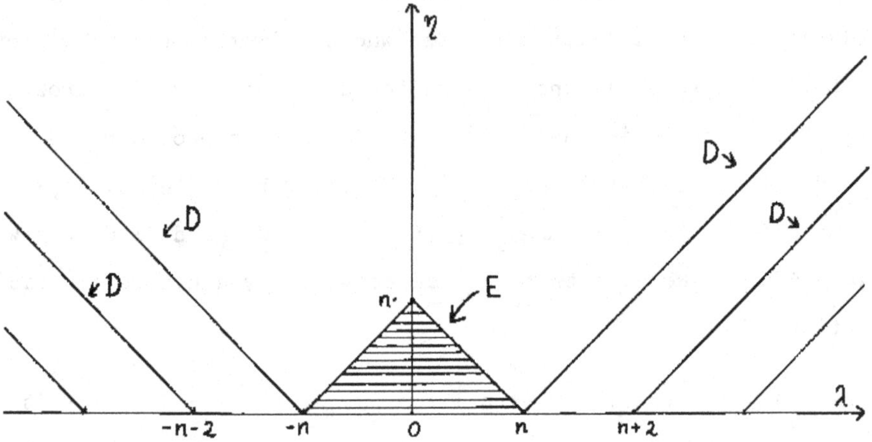

Let $\pi = \pi(\mu,\lambda)$ be the representation corresponding to $\varphi_{(\mu,\lambda)}$. One should think of λ as corresponding to $\delta_\lambda \in K_1^\wedge$ such that $\delta_\lambda \in \pi_{|K}$, and of μ as corresponding to the infinitesimal character of π . For example we read of from the graph that a fixed $\delta \in K_1^\wedge$ can occur in at most finitely many inequivalent representations from the relative discrete series and if $|\lambda| \leqq n$ then δ_λ does not occur in any.

It can be proved that for parametervalues in the triangle $E = \{(\eta,\lambda) \mid |\lambda| + \eta \leqq n\}$, and in the half plane $C = \{(\mu,\lambda) \mid \mu \geqq 0 , \lambda \in \mathbb{R}$. $\varphi_{\mu,\lambda}$ is positive definite. I believe, but I have not proved that for $n > 1$ $D \cup E \cup C$ give all the positive definite spherical functions. (For $n = 1$ the situation is described below). This is what we can say about problem II . For problem III a direct application of the spectral theory of singular differential operators give the Plancherel measure for $SU(n,1)^{\sim}/SU(n)$:

$$|c_1(\mu,\lambda)|^{-2}\, d\mu\, d\lambda \quad \text{over } C ,$$

(3) and
$$\frac{2^{-2n+1}\, \eta\Gamma(n+m)\, \Gamma(\eta+n+m)}{\Gamma(n)^2\, \Gamma(\eta+m+1)\, \Gamma(m+1)}\, d\eta$$

over each line in D .

In particular the set E has measure 0 .

For $n > 1$ problem IV is not interesting, since $SU(n,1) / S(U(n) \times U(1))$ is non-tube type (Corollary 4). For $n = 1$ the diagram and the Plancherel measure above describes the situation for G^1, K^1 , and we must discuss what happens when restricting to $G = SU(1,1)^{\sim}$. First notice that the Plancherel measure for $n = 1$ becomes very simple:

$$\tfrac{1}{4}\mu \cdot \mathrm{Re}[\tanh(\tfrac{\pi}{2}(\mu + i\lambda))\, d\mu\, d\lambda \quad \text{over } C$$

(4)

and
$$\tfrac{1}{2}\eta d\eta (= \tfrac{1}{2}(|\lambda| - 2m - 1)d\lambda) \quad \text{over } D .$$

Notice that if $\lambda_1 = \lambda_2 \pmod{2\mathbb{Z}}$ and (μ,λ_1) and (μ,λ_2) both belong to $C \cup D$, then they occur with the same "weight."

On the other hand if the representations π_1 and π_2 of $SU(1,1)^{\sim}$ corresponding to $\nu_1 = (\mu_1,\lambda_1)$ and $\nu_2 = (\mu_2,\lambda_2)$ are equivalent , then $\mu_1 = \mu_2$, since π_1 and π_2 have the same infinitesimal character, and $\lambda_1 = \lambda_2 \pmod{2\mathbb{Z}}$, since φ_{ν_1} and φ_{ν_2} must agree on $Z(G)$.

Now for a given (μ,λ) to determine which $(\mu,\lambda + 2\nu)$,

$\nu \in \mathbb{Z}$ gives an equivalent representation, is the same as, in the terminology of Bargmann [1] and Pukanszky [9] to determine the spectrum of the operator H_0 in the corresponding representation. In the following table we list the results. We describe an equivalence class of representations by the set of parameters (μ,λ) giving representations in the class (cf. Proposition 6). The equivalence classes are named as in the paper of Pukanszky:

Table 1

$C_q^{(\tau)}$ for $q > \frac{1}{4}$, $0 \leqq \tau < 1$:

$$\mu = \sqrt{4q - 1} \quad , \quad \lambda \in 2\tau + 2\mathbb{Z} .$$

$D_\ell^{\pm}$ for $\ell > 0$:

$$\mu = i|2\ell - 1| \quad , \quad \lambda \in \pm 2\ell \pm 2(\mathbb{N} \cup \{0\})$$

D_0: $\mu = i \quad , \quad \lambda = 0 \quad .$

$E_q^{(\tau)}$ for $0 \leqq \tau < 1$, $\tau(1 - \tau) < q \leqq \frac{1}{4}$:

$$\mu = i\sqrt{1 - 4q} \quad , \quad \lambda \in 2\tau + 2\mathbb{Z} \quad .$$

If we in figure 1 , for $n = 1$, take out one representative for each equivalence class we get the following picture:

Figure 2.

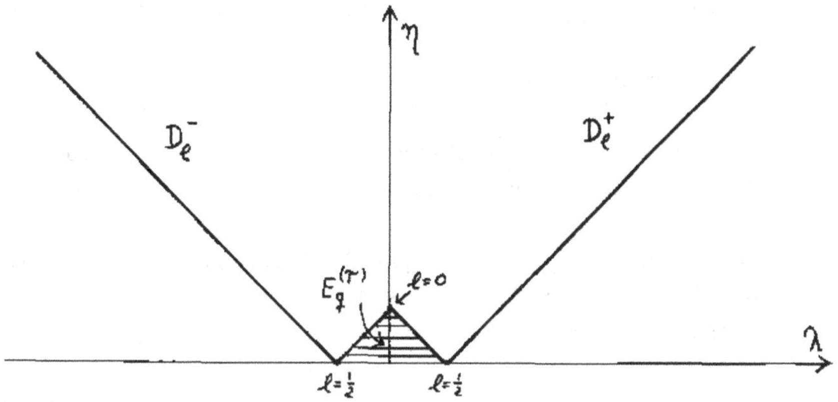

From (4) now follows the Plancherel formula for

$SU(1,1)^\sim$: For all $f \in L^2(SU(1,1)^\sim) \cap L^1(SU(1,1)^\sim)$.

$$\|f\|_2^2 = \frac{1}{4}\int_0^2\int_0^\infty \mathrm{Tr}(\pi_{\mu,\lambda}(f)\ \pi_{\mu,\lambda}(f)^*)\mu \quad \mathrm{Re}[\tanh(\tfrac{\pi}{2}(\mu+i\lambda)]d\mu\ d\lambda\ +$$

$$+\ \tfrac{1}{2}\int_0^\infty \mathrm{Tr}(\pi_{i\eta,\eta+1}(f)\ \pi_{i\eta,\eta+1}(f)^*)\eta\ d\eta$$

$$+\ \tfrac{1}{2}\int_0^\infty \mathrm{Tr}(\pi_{i\eta,-\eta-1}(f)\ \pi_{i\eta,-\eta-1}(f)^*)\eta\ d\eta\ .$$

Where the double integral corresponds to the continuous series and the two single integrals corresponds to the relative discrete series.

References.

[1] Bargmann, V. Irreducible, unitary representations of the Lorentz group.
Ann. of Math. 48(1947), 562-640.

[2] Dixmier, J. Les C^*-algèbres et leurs représentations.
Gauthier-Villars, Paris 1964.

[3] Flensted-Jensen, M. The spherical functions on the universal covering of SU(n-1,1)/SU(n-1).
Preprint, Matematisk Institut, Copenhagen 1973.

[4] Flensted-Jensen, M. Spherical functions on a simply-connected semisimple Lie group.
Preprint, Matematisk Institut, Copenhagen, 1974.

[5] Godement, R. Introduction aux travaux de A. Selberg.
Seminaire Bourbaki, 1957.

[6] Harish-Chandra, Spherical functions on semisimple Lie groups I + II.
Amer. J. Math., 80(1958), 241-310, 553-613.

[7] Harish-Chandra, Discrete series for semisimple Lie groups, II.
Acta Math. 116(1966), 1-111.

[8] Helgason, S. Differential geometry and symmetric spaces.
Academic Press, New York 1962.

[9] Pukanszky, L. The Plancherel formula for the universal covering group of SL($\mathbb{R}$,2).
Math. Ann. 156(1964), 96-143.

Matematisk Institut
Universitetsparken 5,

DK 2100 Kobenhavn
Denmark

GROUPES REDUCTIFS ET GROUPES RESOLUBLES

Paul GERARDIN

1. -

Soit G un groupe algébrique connexe réductif défini sur $\mathbb{R}$ dont le groupe dérivé est simplement connexe. A chaque tore maximal T de G défini sur $\mathbb{R}$ est associée une série de représentations irréductibles de $G(\mathbb{R})$. Cette famille est paramétrée par les orbites des caractères réguliers de $T(\mathbb{R})$ sous le groupe $W(T)$, quotient par $T(\mathbb{R})$ du normalisateur de T dans $G(\mathbb{R})$; la valeur sur les éléments réguliers de $T(\mathbb{R})$ de la trace d'une telle représentation la caractérise entièrement, et c'est ([5]) :

$$s(G,T) \quad \sum_{W(T)} \quad \theta({}^{w}t) \;/\; \Delta_\theta\,({}^{w}t)$$

où $s(G,T)$ est le signe défini par la parité du nombre de racines positives compactes de T relativement au centralisateur de sa partie déployée,

$\Delta_\theta(t)$ est le produit pris sur les racines positives de (G,T) telles que la coracine correspondante soit positive sur la différentielle de θ:

$$\Delta_\theta\,(t) = \prod_{<\theta^0\,,\,\alpha^\vee>>0} \Delta_\alpha(t)$$

$\Delta_\alpha(t)$ étant égal à $t^{\alpha/2} - t^{-\alpha/2}$ si l'action du groupe de Galois Γ de $\mathbb{C}$ sur $\mathbb{R}$ envoie α sur $-\alpha$, et à $|t^{\alpha/2} - t^{-\alpha/2}|$ sinon.

On construit ici des représentations irréductibles du groupe $G(\mathcal{O})$ des points de G à valeurs dans l'anneau $\mathbb{R}[[\Pi]]$ des séries formelles sur $\mathbb{R}$, paramétrées par les caractères réguliers (N° 4) de $T(\mathcal{O})$, ou plus exactement par les orbites de ces caractères par $W(T)$. Un groupe, limite projective de groupes de Lie résolubles joue un rôle essentiel (N° 6). Le principe de la construction est tout à fait analogue à celui utilisé dans [4] ; les résultats de [1], et particulièrement ceux du chapitre IX, ainsi que [3], et [6], ont été utilisés systématiquement.

2. -

Soit G un groupe algébrique connexe réductif défini sur le corps $\underline{R}$ des réels ; on suppose que son groupe dérivé est simplement connexe.

Désignons par $\mathcal{O}$ l'anneau $\underline{R}[[\Pi]]$ des séries formelles à une indéterminée à coefficients réels, et par $\wp$ son idéal maximal. Le corps des restes $\mathcal{O}/\wp$ est donc le corps des réels.

Pour chaque entier $n \geq 1$, l'ensemble des points de G dans l'algèbre $\mathcal{O}/\wp^n$ est un groupe de Lie $G(\mathcal{O}/\wp^n)$. Si $n \geq m$, la réduction

$$\mathcal{O}/\wp^n \longrightarrow \mathcal{O}/\wp^m$$

fournit l'homomorphisme surjectif

$$G(\mathcal{O}/\wp^n) \longrightarrow G(\mathcal{O}/\wp^m) \quad ,$$

dont le noyau, noté $G(\wp^m/\wp^n)$, est un groupe nilpotent. Lorsque $2m \geq n$, ce groupe est commutatif, et s'identifie à $\mathcal{G}(\wp^m/\wp^n)$ si $\mathcal{G}$ désigne l'algèbre de Lie de G. Le groupe $G(\mathcal{O})$ est limite projective des groupes de Lie $G(\mathcal{O}/\wp^n)$ via les applications précédentes. Soit $G(\wp^n)$ le noyau de la réduction modulo $\wp^n$ sur $G(\mathcal{O})$:

$$1 \longrightarrow G(\wp^n) \longrightarrow G(\mathcal{O}) \longrightarrow G(\mathcal{O}/\wp^n) \longrightarrow 1 \quad .$$

Les commutateurs de $G(\wp^n)$ avec $G(\wp^m)$ sont dans $G(\wp^{n+m})$, le quotient $G(\wp^m)/G(\wp^n)$, si $n \geq m$, s'identifie au groupe $G(\wp^m/\wp^n)$. Enfin, si $3m \geq n \geq 2m$, les commutateurs dans le groupe $G(\wp^m/\wp^n)$ sont dans le groupe $G(\wp^{2m}/\wp^n)$ qui s'identifie à l'algèbre de Lie $\mathcal{G}(\wp^{2m}/\wp^n)$, et ils sont donnés par le crochet de Lie des images dans $\mathcal{G}(\wp^m/\wp^{2m})$ via les réductions

$$G(\mathfrak{p}^m/\mathfrak{p}^n) \longrightarrow G(\mathfrak{p}^m/\mathfrak{p}^{2m}) \simeq \mathfrak{g}(\mathfrak{p}^m/\mathfrak{p}^{2m}) ,$$

et

$$\mathfrak{g}(\mathfrak{p}^{2m}/\mathfrak{p}^{3m}) \longrightarrow \mathfrak{g}(\mathfrak{p}^{2m}/\mathfrak{p}^n) \simeq G(\mathfrak{p}^{2m}/\mathfrak{p}^n) .$$

3. -

Soit T un tore maximal de G défini sur $\underline{R}$. On définit les groupes $T(\mathfrak{p}^m)$ et $T(\mathfrak{p}^m/\mathfrak{p}^n)$ comme noyaux des réductions modulo $\mathfrak{p}^m$ sur $T(\mathcal{O})$ et $T(\mathcal{O}/\mathfrak{p}^n)$. On note Γ le groupe de Galois de $\underline{C}$ sur $\underline{R}$; une barre désignera l'action de l'élément non trivial de Γ. Soit R le système de racines de (G, T). Le groupe Γ opère sur R ; pour chaque orbite du groupe $\pm\Gamma$ dans R, soit Ω, on définit un tore T_Ω : il est engendré par l'image des coracines $\alpha^\vee$ et $\bar{\alpha}^\vee$ avec $\alpha \in \Omega$ (il ne dépend pas du choix de α dans Ω) .

4. -

Soit θ un caractère continu de $T(\mathcal{O})$. Pour chaque orbite $\Omega \subset R/\pm\Gamma$, on note $|\Omega|_\theta$ le plus petit entier n tel que la restriction de θ à $T_\Omega(\mathfrak{p}^n)$ soit triviale. On dit que le caractère θ est <u>régulier</u> si on a

$$|\Omega|_\theta \geq 1 , \text{ pour tout } \Omega \in R/\pm\Gamma .$$

On écrit $|\alpha|_\theta$ pour $|\Omega|_\theta$ lorsque $\alpha \in \Omega$. Si θ est un caractère régulier de $T(\mathcal{O})$, et si $\alpha, \beta, \gamma \in R$ et $\alpha + \beta + \gamma = 0$, on a :

$$|\alpha|_\theta \leq \operatorname{Max}(|\beta|_\theta , |\gamma|_\theta)$$

Soit θ un caractère régulier de $T(\mathcal{O})$. Pour chaque entier $i \geq 1$, les racines $\alpha \in R$ telles que $|\alpha|_\theta \leq i$ forment un sous-système de racines R_i de R ; on définit alors un sous-groupe H_i de $G(\mathcal{O})$ de la façon suivante :

- pour $i = 1$, H_1 est le groupe $G^{R_1}(\mathcal{O})$, points rationnels sur $\mathcal{O}$ du groupe algébrique réductif défini sur $\underline{R}$ déterminé par le tore maximal T et le système de racines R_1, qui est invariant par Γ ;

- soit m_i le sous-espace vectoriel de $\mathfrak{g}$ engendré par les sous-espaces radiciels relatifs aux racines α de $|\alpha|_\theta = i$; désignons par G^{R_i} le groupe algébrique réductif que définissent T et R_i munis de l'action de Γ ; soit M_i

l'image réciproque dans $G^{R_i}(\mathfrak{p}^{i'})$, où i' désigne la partie entière de $i/2$, du sous-espace $m_i(\mathfrak{p}^{i'}/\mathfrak{p}^{2i'})$ de

$\mathfrak{g}^{R_i}(\mathfrak{p}^{i'}/\mathfrak{p}^{2i'})$ via la réduction modulo $\mathfrak{p}^{2i'}$ sur

$$G^{R_i}(\mathfrak{p}^{i'}) \longrightarrow G^{R_i}(\mathfrak{p}^{i'}/\mathfrak{p}^{2i'}) \simeq \mathfrak{g}^{R_i}(\mathfrak{p}^{i'}/\mathfrak{p}^{2i'}) \quad ;$$

si donc $i > 1$, le groupe H_i est le sous-groupe de $G(\mathcal{O})$ qu'engendrent $T_\Omega(\mathfrak{p}^{2i'})$ pour $|\Omega|_\theta = i$, et M_i ; c'est une limite projective de groupes nilpotents, et on a une surjection :

$$H_i \longrightarrow m_i(\mathfrak{p}^{i'}/\mathfrak{p}^{i''}) \quad \text{où } i'' = i - i' \quad .$$

Les commutateurs de H_i avec H_j sont dans H_{i+j} . L'ensemble $\prod_{i \geq 2} H_i$, noté H^1, est une limite projective de groupes nilpotents, normalisé par $T(\mathcal{O})$ et H_1. Le groupe $T(\mathcal{O})H^1$ est une limite projective de groupes résolubles.

5. -

Pour chaque $i \geq 1$, on définit une représentation irréductible η_i de H_i :

- Si $i = 1$, c'est la représentation de H_1 fournie par les résultats d'Harish-Chandra : le caractère θ définit un caractère régulier du tore maximal du groupe dérivé de $G^{R_1}(\underline{\underline{R}})$ et donc une représentation irréductible de ce groupe ; on en déduit une représentation de H_1 en prolongeant cette représentation par l'homothétie que définit θ sur le sous-groupe qu'engendrent les $T_\Omega(\mathcal{O})$ pour $|\Omega|_\theta > 1$;

- Si $i > 1$ est pair, η_i est la représentation unité ;

- Si $i > 1$ est impair, η_i est la représentation de H_i égale à θ sur le sous-groupe $T_i(\mathfrak{p}^{2i'})$ qu'engendrent les $T_\Omega(\mathfrak{p}^{2i'})$ pour $|\Omega|_\theta = i$, et triviale sur l'image réciproque de $m_i(\mathfrak{p}^{i''}/\mathfrak{p}^{i})$ par

$$H_i \cap G^{R_i}(\mathfrak{p}^{i''}) \to G^{R_i}(\mathfrak{p}^{i''}/\mathfrak{p}^{i}) \simeq \mathfrak{g}^{R_i}(\mathfrak{p}^{i''}/\mathfrak{p}^{i}) \supset m_i(\mathfrak{p}^{i''}/\mathfrak{p}^{i})$$

6. -

Le groupe H_1 conserve la classe de la représentation η_i. Lorsque i est pair, on prolonge trivialement η_i à $H_1 H_i$; lorsque i est impair > 1, la représentation η_i provenant de la représentation égale à l'identité sur le groupe $\underline{\underline{U}}_1$ des nombres complexes de module 1 via la surjection

$$H_i \longrightarrow \underline{\underline{U}}_1 \, . \; m_i(\mathfrak{p}^{i'}/\mathfrak{p}^{i''})$$

définie par le caractère θ et la réduction modulo $\mathfrak{p}^{i''}$, admet une réalisation "à la Schrödinger", et le cocycle de la représentation correspondante, projective, de $T(\mathcal{O})$ se calcule grâce aux formules explicites que donne A. WEIL ; ce cocycle est trivial.

7. -

Fixons un demi-espace ouvert de l'espace des racines dont R_1 soit la trace sur R de son bord ; les racines qu'il contient forment une part horicyclique R^{1+} invariante par R_1 ; si η^1 est la représentation de H^1 que définit le produit tensoriel des η_i relativement aux intersections des H_i entre eux, on en déduit une polarisation positive relativement à la représentation η^1 qui est invariante par l'action de H_1, ce qui permet de prolonger η^1 à H_1 ; en prenant son produit tensoriel avec η_1, on construit une représentation irréductible de $H_1 H^1$, soit η_θ. Lorsque $t \in T(\underline{\underline{R}})$ est un élément régulier, l'opérateur $\eta_\theta(t)$ est traçable, et sa trace est donnée par :

$$\mathrm{Tr}\, \eta_\theta(t) = s(G, T) \sum_{W_1(T)} \theta({}^{w}t) / \Delta_\theta({}^{w}t)$$

où $W_1(T)$ est le quotient par $T(\underline{B})$ de son normalisateur dans $G^{R_1}(\underline{B})$,

$s(G, T)$ est le signe donné au N° 1 ,

$\Delta_\theta(t) = \prod_{<\theta_\Omega, \alpha^\vee>>0} \Delta_\alpha(t)$, où, pour chaque racine $\alpha \in R$, d'orbite Ω par $\pm\Gamma$, on note θ_Ω l'élément $t'_\Omega(\underline{B})$ que définit la différentielle du caractère de $T_\Omega(\wp^{i-1}/\wp^i)$, $|\Omega|_\theta = i$,

ceci en choisissant convenablement le prolongement de η^1 à H_1 .

8. -

La représentation de $G(\mathcal{O})$ qu'induit la représentation η_θ de H_1H^1 est irréductible, et, sur les éléments réguliers de $T(\underline{B})$, les opérateurs de la représentation ont une trace égale à

$$s(G, T) \sum_{W(T)} \theta({}^w t)/\Delta_\theta({}^w t) \quad .$$

9. -

Suivant le même principe, on peut construire par cette méthode des représentations irréductibles des groupes $\wp$ -adiques ; lorsque le tore maximal correspondant est non ramifié minisotrope et spécial, on obtient de cette manière des représentations surcuspidales relatives aux caractères réguliers de ce tore ; lorsque R_1 n'est pas vide, il faut utiliser la conjecture de Macdonald ([2], 6. 7). Un certain nombre de caractéristiques résiduelles doivent être éliminées pour avoir une formule "universelle". Un cas particulier est traité dans [4] .

REFERENCES.

[1] BERNAT, P., CONZE, N., DUFLO, M., LEVY-NAHAS, M., RAIS, M., RENOUARD, P., VERGNE, M. : Représentations des groupes de Lie résolubles. Monographies de la Société Mathématique de France, 4, Paris, Dunod 1972.

[2] BOREL, A., CARTER, R., CURTIS, C.W., IWAHORI, N., SPRINGER, T.A., STEINBERG, R. : Seminar on algebraic groups and related finite groups. Lecture Notes in Mathematics, 131, Berlin, Springer-Verlag (1970).

[3] DUFLO, M. : Sur les extensions des représentations irréductibles des groupes de Lie nilpotents. Ann. Sc. de l'E. N. S., 5, 71 - 120 (1972).

[4] GERARDIN, P. : Sur les séries discrètes non ramifiées des groupes réductifs déployés $\mathfrak{p}$-adiques. Thèse, Paris (juin 1974).

[5] HARISH-CHANDRA : Harmonic analysis on semi-simple Lie groups. Bull. Amer. Math. Soc. 76, 529-551 (1970).

[6] WEIL, A. : Sur certains groupes d'opérateurs unitaires. Acta Math. 111, 143-211 (1964).

Université de Paris VII
U. E. R. de Mathématiques
2, place Jussieu

75221 PARIS CEDEX 05

STABILITY AND EQUILIBRIUM IN QUANTUM STATISTICAL MECHANICS

Daniel KASTLER

This is a report on a common work with Rudolf Haag and Eva Trych-Pohlmeyer, which is technically connected with the harmonic analysis of non-commutative dynamical systems. The general aim of this work is to provide a derivation of the Gibbs Ansatz, base of the equilibrium Statistical Mechanics, from a stability requirement. By the same token a relation is established between stability and the positivity of the hamiltonian in the zero temperature case.

Rather than the Gibbs Ansatz pertaining to finite systems, we derive in fact the so-called Kubo-Martin-Schwinger (K.M.S.) condition, a substitute of the Gibbs Ansatz for infinite quantum systems. Since these concepts are not generally familiar to functional analysts (although the second now plays a central role in the theory of Von Neumann Algebras) we shall first describe them by sketching the way in which temperature equilibrium states are obtained mathematically in standard Quantum Statistical Mechanics. This is done in two steps :
1) One first considers finite systems i.e., physically, systems describing a finite portion of the substance under consideration enclosed in a cubic box of length L, with appropriate boundary conditions on the walls of the box. Mathematically this has two consequences :

- first, the algebra of observables can be chosen to be the algebra $\mathfrak{A}$ of the compact operators on some Hilbert space $\mathcal{H}$;
- second, the dynamical evolution of the system is described by a positive self-adjoint operator H with a pure point spectrum(1), each point having a finite multiplicity, so that $e^{-\beta H}$ is trace class for each positive β. This Hamiltonian H induces a one-parameter group $t \to \alpha_t$ of automorphisms of $\mathfrak{A}$ in the following way : for $A \in \mathfrak{A}$

$$(1) \qquad \alpha_t(A) = e^{itH} A e^{-itH} .$$

The Gibbs Ansatz for describing the state ω of the finite system corresponding to the inverse temperature β then consists in assuming that

(1) general feature of the energy eigenvalue problem for a finite quantum system in a box.

(2) $$\omega(A) = \frac{\mathrm{Tr}\{e^{-\beta H} A\}}{\mathrm{Tr}\{e^{-\beta H}\}}$$

This formula defines a state (normalized positive functional) of $\mathfrak{A}$, whose physical meaning is that $\omega(A)$ is the mean value of the observable A in the physical state corresponding to the inverse temperature β, a description in accordance with the manner in which Quantum Mechanics describes physical states. number of particles proportional to its volume. The existence of one or several limit states is then guaranteed by the analytical properties of the hamiltonian, as can be proved with sufficient ingenuity and mathematical skill (the proof has been given for a number of models - the investigation of this "thermodynamical limit" is one of the principal aims of what one might call "Constructive Statistical Mechanics", which is not our concern in this work).

What is, now, the K.M.S. condition? We obtain it in the following way : notice that, for an $A \in \mathfrak{A}$ which is an analytic vector for the one-parameter group (1) one can extend α to an imaginary value $i\beta$ of time :

$$\alpha_{i\beta}(A)^{\bullet} = e^{-\beta H} A\, e^{\beta H}$$

One has then, for all $B \in \mathfrak{A}$, from (2),

$$\omega(B\alpha_{i\beta}(A)) = \frac{\mathrm{Tr}\{e^{-\beta H} B\, e^{-\beta H} A\, e^{\beta H}\}}{\mathrm{Tr}\{e^{-\beta H}\}} = \frac{\mathrm{Tr}\{e^{-\beta H} BA\}}{\mathrm{Tr}\{e^{-\beta H}\}}$$

Whence the <u>K.M.S. condition</u> :

(3) $$\omega(B\alpha_{i\beta}(A)) = \omega(AB)$$

This condition (3) has two advantages : first it persists in the thermodynamic limit (as can be checked on various models) and thus affords a substitute of (2) valid for infinite systems (for which (2) itself makes no sense), substitute which contains, as our experience shows, the same amount of information as (2). Thus we can replace the Gibbs Ansatz (2) by the K.M.S. condition (3) of more general validity, yielding the fundament of the Quantum Statistical Mechanics of infinite systems rather than of finite ones. This is important in that the infinite systems are the ones whose features correspond to a "thermodynamical behaviour" (which does not show up in finite systems- this motivates the necessity of performing the thermodynamic limit in the

traditional approach described above). Physically: a system containing 10^{23} particles is best idealized by considering an infinite number of particles.
A second advantage of the K.M.S. condition (3) is that it has become one of the central items in the theory of Von Neumann algebras, and is therefore, mathematically, a beautiful object. We close this discussion of traditional Quantum Statistical Mechanics by noting that, if we introduce the functions

$$(4) \qquad \begin{cases} F_{AB}(t) = \omega(B\alpha_t(A)) - \omega(A)\,\omega(B) & A,B \in \mathfrak{A} \\ G_{AB}(t) = \omega(\alpha_t(A)B) - \omega(A)\,\omega(B) & t \in R \end{cases}$$

(3) can be written equivalently

$$(5) \qquad \hat{F}_{AB}(E) = e^{\beta E}\,\hat{G}_{AB}(E) \qquad , A,B \in \mathfrak{A}$$

in terms of the Fourier transforms $\hat{F}_{AB}$ and $\hat{G}_{AB}$ of the functions (4) (if we assume for the automorphism group α the natural continuity property that $t \in R \to \varphi(\alpha_t(A))$ should be continuous for all $A \in \mathfrak{A}$ and all states φ of $\mathfrak{A}$, $\hat{F}_{AB}$ and $\hat{G}_{AB}$ are bounded measures, for which the condition (5) has been written in a somewhat sloppy way as if these measures were functions of E (the energy), which will in fact be the case owing to further assumptions). The reason why we mention the alternative (5) to the classical K.M.S. condition (3) is that (5) naturally lends itself to our proof (note, also, that (5) can be stated without the restriction that A be an analytic element of the one-parameter group α).

We now turn to our objective, which is to give the K.M.S condition (5), the status of a theorem rather than that of an Ansatz, starting from scratch. For this we consider, from the start, an infinite quantal system, which we idealize as a "<u>C^* - system</u>",i.e. a pair $\{\mathfrak{A},\alpha\}$ of a C^*-algebra $\mathfrak{A}$ and a continuous one-parameter group α of automorphisms of $\mathfrak{A}$ (the continuity assumption is the natural one that all numerical functions $t \in R \to \varphi(\alpha_t(A))$, $A \in \mathfrak{A}$ φ a state of $\mathfrak{A}$, shall be continuous -one could require, equivalently, that the map $t \in R \to \alpha_t(A)$ be continuous for all $A \in \mathfrak{A}$). This notion of C^*-system is a mathematical abstraction of the general frame of quantum mechanics for the description of a physical system together with its time evolution (= dynamics). It is both relevant for finite systems (in which case $\mathfrak{A}$ can be chosen as the algebra of compact operators on some Hilbert space) and for infinite systems : then $\mathfrak{A}$ is an "antiliminar" C^*-algebra (= possessing a maze of inequivalent representations) whose complexity reflects that of infinite system. The elements of $\mathfrak{A}$ represent physically (norm limits of) local observables, whereby $\alpha_t(A)$, $A \in \mathfrak{A}$, $t \in R$, represents the observable obtained from A by a shift t in time. Physical states are defined by the states (= normalized positive functionals) of $\mathfrak{A}$, the value $\omega(A)$ of the state φ for $A \in \mathfrak{A}$ representing the mean value of the observable A in the state φ . Our program is now to derive the K.M.S. condition (5) (or positivity of the energy, the limiting case of (5) for zero temperature) from physically natural requirements on equilibrium states. The three constitutive properties required for an equilibrium state ω are the following :

(i) ω is invariant under α ;

(ii) ω is an extremal element of the convex set of α-invariant states ;

(iii) ω is stable for local perturbations of the dynamics.

Before starting with our argument we briefly comment upon these conditions. First, from a physical point of view, it should be clear that (i), (ii) and (iii) are natural requirements for characterizing thermodynamical equilibrium states : (i) is obvious; (ii) corresponds to the fact that we want to describe "pure thermodynamical phases" rather than quantal mixtures of them; and, as for (iii), its physical meaning is clear; a local disturbance of the dynamics (e.g., an impurity in a crystal, a boat on an ocean) should not upset the original state, but merely cause a gentle distortion. Our second comment is that, of course, the conditions (i), (ii), (iii) above have to be stated mathematically in a precise way. For (i) this is obviously done as follows :

<u>Assumption (i)</u> : (invariance)

(6) $$\omega(\alpha_t(A)) = \omega(A) \qquad \text{for all } A \in \mathfrak{A} \text{ and } t \in R$$

Conditions (ii) and (iii), on the other hand, will be given precise formulations as we need them in our proof (in fact the technical conditions which we will need will turn out to be somewhat strong mathematical exegeses of (ii) and (iii) as phrased above, which we hope future progress will help to release).

Now down to work! Since the first part of our argument consists in exploiting (iii) in combination with (6), we now need to formulate (iii). For this we need a mathematical formulation of "local perturbations of the dynamics". That is done as follows :

consider $h = h^* \in \mathfrak{A}$ and define the differentiable function

(7) $t \in R \to P_t^{(h)} \in \mathfrak{A}$

by the following differential equation and boundary condition at t=0

$$(8)\quad \begin{cases} i\,\dfrac{dP_t^{(h)}}{dt} = P_t^{(h)}\,\alpha_t(h) \\ P_0^{(h)} = I \end{cases}$$

(these entail the existence and uniqueness of the function (7)). One easily shows that the solution $P^{(h)}$ of (8) is a unitary cocycle in the following sense :

$$(9)\quad \begin{cases} P_t^{(h)*} = P_t^{(h)^{-1}} = \alpha_t(P_{-t}^{(h)}) \\ P_{s+t}^{(h)} = P_s^{(h)}\,\alpha_s(P_t^{(h)}) \end{cases} \qquad ,s,t \in R ,$$

allowing the definition

$$(10)\quad \alpha_t^{(h)}(A) = P_t^{(h)}\alpha_t(A)\,P_t^{(h)*} \qquad , A \in \mathfrak{A},\ t \in R ,$$

of a "perturbed one-parameter group" $\alpha_t^{(h)}$, depending upon the choice of the self-adjoint $h \in \mathfrak{A}$. The fact that we have, here, a description of a "local perturbation of the dynamics" stems from the property[1]

$$(11)\quad \left.\frac{d}{dt}\right|_{t=0} \alpha_t^{(h)}(B) = i\left.\frac{d}{dt}\right|_{t=0} \alpha_t(B) + [h,B]$$

easily derived from (8) for a "differentiable" $B \in \mathfrak{A}$ (one for which $t \to \alpha_t(B)$ is differentiable) : (11) shows that in a representation of $\mathfrak{A}$ where α is obtained from a hamiltonian H as in (1), $\alpha^{(h)}$ is likewise obtained from the hamiltonian $H \dotplus h$, a local dynamical perturbation since h represents a local observable. Equipped with this description of local perturbations[2] we are now ready to formulate precisely

---

(1) [ ] denotes a commutator.

(2) which we owe to Derek Robinson and Huzihiro Araki (and which constitutes basically a bounded operator version of the old Dirac-Tomonaga-Dyson perturbation expansion).

Assumption (iii) : (stability)

For each self-adjoint h in $\mathfrak{A}$ there is a map $\lambda \to \omega^{(\lambda h)}$ of a neighbourhood $\mathcal{V}_h$ of zero in R to the state space of $\mathfrak{A}$ such that :

a) $\omega^{(\lambda h)}$ is invariant for the perturbed dynamics $\alpha^{(\lambda h)}$:

$$(12) \qquad \omega^{(\lambda h)}(\alpha_t^{(\lambda h)}(A)) = \omega^{(\lambda h)}(A) \qquad , \quad t \in R \quad , \quad A \in \mathfrak{A},$$

b) $\lambda \to \omega^{(\lambda h)}$ is differentiable, in the weak sense, for $\lambda = 0$, with derivative $\omega_1^{(h)}$:

$$(13) \qquad \left.\frac{d}{d\lambda}\right|_{\lambda=0} \omega^{(\lambda h)}(A) = \omega_1^{(h)}(A) \qquad , \quad A \in \mathfrak{A},$$

c) $\omega_1^{(h)}$ is a normal form of the representation π_ω of $\mathfrak{A}$ generated by ω[1] .

From this stability assumption a very simple argument allows to proceed towards our aim of proving condition (5) : from (12) immediately follows that, for each differentiable $B \in \mathfrak{A}$,

$$0 = \omega^{(\lambda h)}\left(\left.\frac{d}{dt}\right|_{t=0} \alpha_t^{(\lambda h)}(B)\right) \qquad ,$$

whence, using (11),

$$0 = \omega^{(\lambda h)}\left(i \left.\frac{d}{dt}\right|_{t=0} \alpha_t(B) + [\lambda h, B]\right) .$$

Replacing $\omega^{(\lambda h)}$ by $\omega^{(\lambda h)} - \omega + \omega$ and dividing by λ , one obtains, taking account of (6) differentiated with respect to t for $t=0$

$$0 = \omega([h,B]) + \frac{\omega^{(\lambda h)} - \omega}{\lambda}\left(i \left.\frac{d}{dt}\right|_{t=0} \alpha_t(B) + \lambda[h,B]\right) \quad ,$$

whence, by (13), for $\lambda \to 0$,

$$(14) \qquad 0 = i\omega_1^{(h)}\left(\left.\frac{d}{dt}\right|_{t=0} \alpha_t(B)\right) + \omega([h,B]) \quad .$$

If, in this equation, we take B to be the differentiable

$$B = \int_S^T \alpha_t(A)\, dt \qquad , \quad A \in \mathfrak{A} \ , \ S,T \in R$$

---

(1) c) is not strictly necessary for our proof and is perhaps too strong as formulation. Cf. discussion below.

with (as immediately checked)

$$i \frac{d}{dt}\Big|_{t=0} \alpha_t(B) = \alpha_T(A) - \alpha_S(A) \quad ,$$

we obtain that, for all $S,T \in R$,

(15) $$\omega_1^{(h)}(\alpha_T(A) - \alpha_S(A)) = i \int_S^T \omega([h, \alpha_t(A)])\, dt \quad .$$

We have not yet used condition c) in Assumption (iii) above : c) will be used in combination with a (strengthened form of) the extremality assumption (ii) which we formulate as follows :

Assumption (ii) : (a strengthened form of the extremality of the invariant state ω).

We assume that :

a) the C^*-system $\{\mathfrak{A}, \alpha\}$ is asymptotically abelian in the sense that, for any $A,B \in \mathfrak{A}$, and state φ of $\mathfrak{A}$,

(16) $$\varphi([A,\alpha_t(B)]) \xrightarrow[t=\infty]{} 0$$

b) the state ω is hyperclustering in the following sense : there is a dense, self-adjoint set $\mathcal{S}$ of $\mathfrak{A}$ such that, for arbitrary $A,B \in \mathcal{S}$ there is a majoration

(17) $$|\omega(A\alpha_t(B)) - \omega(A)\ \omega(B)| < \frac{C}{\{1 + |t|\}^{1+\delta}} \quad ,$$

where C and δ are positive constants.

Furthermore analogous majorations hold for the truncated expectation values up to order 6[1] .

Why is this assumption a stronger form of extremality for ω ? Because, for asymptotically abelian systems, we have the fact that extremality of an α-invariant state ω is synonimous with the fact that , for $A,B \in \mathfrak{A}$,

(18) $$\omega(A\alpha_t(B)) \xrightarrow[t=\infty]{} \omega(A)\ \omega(B) \qquad \text{in mean.}$$

Physical reasons for strengthening this latter condition can be given (work in progress on this point). We uncritically adopt the above formulation of Assumption (ii) to conclude that, for $A,B \in \mathfrak{A}$,

---

(1) We refrain from stating these conditions precisely, because they are used in a later stage of our proof, which will only be sketched here. For a precise proof the reader is referred to [9].

$$(19) \qquad \int_{-\infty}^{+\infty} \omega\,([h,\alpha_t(A)])\,dt = 0 \qquad ,$$

(with independent limits in the integration). This conclusion follows from performing the limits $S \to -\infty$, $T \to +\infty$ in (15), noting that (18) (a fortiori (19)), combined with asymptotic abelianness, entails that

$$\alpha_t(A) \xrightarrow[t=\pm\infty]{} \omega(A)\, I \qquad ,$$

σ -weakly, in the representation generated by the state ω , whence the vanishing of the l.h.s. of (15) in the limit. It is now apparent that we achieved progress towards proving the K.M.S. condition (5) since (19) can alternatively be written

$$(20) \qquad \int_{-\infty}^{+\infty} F_{Ah}(t)\,dt = \int_{-\infty}^{+\infty} G_{Ah}(t)\,dt$$

(Cf. (4)) , which is nothing but the special case of (5) for $E=0$.

The rest of our work consists in having the "K.M.S. condition at zero energy" (20) to finite values of the energy. This is done by using the following trick (we here sketch the proof, omitting details needed for rigour) : since (20) holds for arbitrary $h,A \in \mathfrak{A}$, it is liable to make $h=h_1\alpha_u(h_2)$, $A = A_1\alpha_u(A_2)$ with $h_1,h_2,A_1,A_2 \in \mathcal{S}$ and $u \in R$. Using the fact that, by the hyperclustering property of ω ,

$$\omega(h_1\alpha_u(h_2)\,\alpha_t(A_1)\,\alpha_u\,(\alpha_t(A_2))) \xrightarrow[t=\infty]{} \omega(h_1\alpha_t(A_1))\,\omega(h_2\alpha_t(A_2)) \quad ,$$

and taking (allowable) limits under the integral, one derives that

$$(21) \qquad \int_{-\infty}^{+\infty} F_{A_1h_1}(t)\,F_{A_2h_2}(t)\,dt = \int_{-\infty}^{+\infty} G_{A_1h_1}(t)\,G_{A_2h_2}(t)\,dt \quad ,$$

$$h_1,\ h_2,\ A_1,\ A_2 \in \mathcal{S};$$

and also, by an iteration of the same trick, that

$$(22) \qquad \int_{-\infty}^{+\infty} F_{A_1h_1}(t)\,F_{A_2h_2}(t)\,F_{A_3h_3}(t)\,dt =$$

$$\int_{-\infty}^{+\infty} G_{A_1h_1}(t)\,G_{A_2h_2}(t)\,G_{A_3h_3}(t)\,dt \quad ,$$

$$h_1,\ h_2,\ h_3,\ A_1,\ A_2,\ A_3 \in \mathcal{S}.$$

Now, from (21) written with $\alpha_s(A_2)$ instead of A_2, integration of both sides with respect to s after multiplication by e^{-iEs} yields that

$$(23) \qquad \hat{F}_{A_1h_1}(-E)\,\hat{F}_{A_2h_2}(E) = \hat{G}_{A_1h_1}(-E)\,\hat{G}_{A_2h_2}(E)$$

or, using the evident fact

$$(24) \qquad \hat{F}_{Ah}(-E) = \hat{G}_{hA}(E) \qquad ,$$

that

$$(25) \qquad \hat{G}_{h_1A_1}(E)\,\hat{F}_{A_2h_2}(E) = \hat{F}_{h_1A_1}(E)\,\hat{G}_{A_2h_2}(E) \; .$$

this means already, in view of the arbitrariness of choice of h_1, h_2, A_1, A_2, in $\mathcal{S}$, that we have a universal function $\phi(E)$ for which

$$(26) \qquad \hat{F}_{Ah}(E) = \phi(E)\,\hat{G}_{Ah}(E) \qquad , A,h \in \mathfrak{A},$$

provided we have the guarantee that h_1 and A_1 can be chosen so that $\hat{G}_{h_1A_1}(E) \neq 0$ for a preassigned $E \in R$. Assuming this for a while, we note that (26) is identical with (5) if $\phi(E) = e^{-\beta E}$. That the latter is the case now follows from the positivity of ϕ (due to the positivity of $\hat{F}_{AA^*}$ and $\hat{G}_{AA^*}$), and its multiplicity :

$$(27) \qquad \phi(E' + E'') = \phi(E')\,\phi(E'') \quad ,$$

which itself follows from

$$(28) \qquad \phi(-E) = \phi(E)^{-1}$$

(immediate consequence of (24)), combined with (22) exploited in a manner analogous to the step of passing from (21) to (23) . We thus proved (5) if we know that, to each $E \in R$, there is a choice of A_1, $h_1 \in \mathfrak{A}$ with $\hat{G}_{h_1A_1}(E) \neq 0$. This restriction is now settled by the

<u>Proposition</u>

Let $\{\pi, U\}$ be the covariant representation of the C^*-system $\{\mathfrak{A}, \alpha\}$ generated by an invariant state ω of $\mathfrak{A}$ and assume that Assumption (ii) above holds. Then the spectrum Sp(U) of the representation U of R is either one-sided (= lies on the non-negative or the non-positive reals) or coincide with the whole real line R .

The alternative stated by this proposition allows to conclude that if Sp(U) is not one-sided, then Sp(U) = R , whence the possibility of choosing, to each $E \in R$, $A_1 = h_1^* \in \mathfrak{A}$ with $\hat{G}_{A_1^*A_1}(E) \neq 0$. We thus have the

<u>Theorem</u>

Let $\{ \mathfrak{A}, \alpha \}$ be a C^*-system, with ω a state of $\mathfrak{A}$ satisfying assumptions (i), (ii) and (iii) above. Either the spectrum of Sp(U) (defined in the preceeding Proposition) is one-sided, or ω fulfills the K.M.S. condition (5) for some real temperature β .

We know sketch the proof of the Proposition. For $A \in \mathfrak{A}$ we denote by $Sp^\alpha(A)$ the support of the operator valued distribution $\hat{X}_A$, Fourier transform of the function $X_A : t \in R \to \alpha_t(A) \in \mathfrak{A}$. Since $X_{AB} = X_A X_B$, and Fourier transforms turn products into convolutions, it is intuitive that

$Sp^\alpha(AB) \subset Sp^\alpha(A) + Sp^\alpha(B)$ for $A,B \in \mathfrak{A}$. Now with π, U and Ω the G.N.S. construction afforded by the state ω , i.e.

$$(29) \quad \begin{cases} \omega(A) = (\Omega, \pi(A)\Omega) \\ \pi(\alpha_t(A)) = U_t \; \pi(A) \; U_t^* \\ U_t\Omega = \Omega \end{cases} \qquad , t \in R \quad , A \in \mathfrak{A} \; ,$$

One easily checks that $\lambda \in \hat{R}$ is contained in Sp(U) iff, to each neighbourhood $\mathcal{V}$ of λ , there is $A \in \mathfrak{A}$ with $Sp^\alpha(A) \subset \mathcal{V}$ and $\pi(A)\Omega \neq 0$ (to establish this, use that $\mu \in Sp^\alpha(A)$ iff $\hat{f}(\lambda) = 0$ for each $f \in L^1(R)$ with $\alpha_f(A) = 0$)[1]. Now the first step in proving our Proposition consists in establishing that Sp(U) is additive, which goes as follows : given $\lambda_1, \lambda_2 \in$ Sp(U) and an arbitrary neighbourhood $\mathcal{V}$ of $\lambda_1 + \lambda_2$ there are neighbourhoods $\mathcal{V}_1, \mathcal{V}_2$ resp. of λ_1, λ_2 with $\mathcal{V}_1 + \mathcal{V}_2 \subset \mathcal{V}$; and elements A_1, A_2 of $\mathfrak{A}$ with $Sp^\alpha(A_i) \subset \mathcal{V}_i$ and $\pi(A_i)\Omega \neq 0$, i=1,2 . Let $A = \alpha_t(A_1) A_2$, one has

$$Sp^\alpha(A) \subset Sp^\alpha(\alpha_t(A_1)) + Sp^\alpha(A_2) \subset \mathcal{V}_1 + \mathcal{V}_2 \subset \mathcal{V}$$

for each $t \in R$ (observe that $Sp^\alpha(\alpha_t(A_1)) = Sp^\alpha(A_1)$); and t can be chosen such as to make $\pi(A)\Omega \neq 0$ since

---

(1)

$$\alpha_f(A) = \int f(t) \; \alpha_t(A) \; dt$$

$$\|\pi(A)\Omega\|^2 = \omega(A_2^*\alpha_t(A_1^*A_1)A_2) \xrightarrow[t=\infty]{}$$

$$\omega(A_2^*A_2)\,\omega(A_1^*A_1) = \|\pi(A_2)\Omega\|^2\,\|\pi(A_1)\Omega\|^2 ,$$

by the assumed clustering (iii) b) of ω . We thus conclude that $\lambda_1 + \lambda_2 \in Sp(U)$ Sp(U) is additive.

Rest the proof of the Proposition : if Sp(U) is not one-sided it contains $a > 0$ and $-b$, $b > 0$. If a and b are not commensurable, the set of $ma-nb$, m, n are positive integers, will have zero set distance to 0 , whence the density of Sp(U) in R , whence Sp(U) = R since Sp(U) is closed. If a and b are commensurable, one can use the fact that Sp(U) has no isolated points (an easy consequence of Assumption (ii) to replace b by b' , $b' > 0$ not commensurable with a ; and to argue as above.

We conclude with a few remarks. First, our Theorem is satisfactory from a physical point of view, since the alternative of one-sidedness of Sp(U) or K.M.S. nature of ω is what we observed in nature where the first case occurs at zero temperature (where the hamiltonian is known to be positive) and the second for a finite temperature. However what is observed is the positivity of and the occurence of K.M.S. for positive values of β . This is not explained by our work as it stands now and presents us with one of our future problems.

Second; the reader of books on Statistical Mechanics finds that what is observed is the validity of the Gibbs Ansatz (or, for that matter, K.M.S.) with the hamiltonian H replaced by $H - \mu N$, where N is the particle number operator (generator of the gauge group) and μ the chemical potential. This result is obtained by our method replacing the algebra $\mathfrak{A}$ of observables by the field algebra $\mathcal{F}$ and looking for the α-invariant, hyperclustering states of $\mathcal{F}$ stable for local perturbations $\alpha^{(h)}$ of α corresponding to a gauge invariant h . This theory generalizes in fact to arbitrary (non commutative) compact automorphism group commuting with the dynamical group α . Work in collaboration with Rudolf Haag on this subject is in progress.

We conclude with a sketch of an alternative technique for deriving K.M.S. from stability, within a frame less interesting for physics but more in the mood of operator theory. Apart from a possible intrinsic interest for the theory of Von Neumann algebras, this alternative approach has the merit of shedding more light on the mathematical mechanism linking modular automorphisms with stability. Consider a W*-system i.e. a pair $\{ \mathcal{M} , \alpha \}$ of a Von Neumann algebra $\mathcal{M}$ with a one-parameter group α of automorphisms of $\mathcal{M}$ such that $t \in R \to \varphi(\alpha_t(A))$ is continuous for all $A \in \mathcal{M}$ and all normal states φ of $\mathcal{M}$. And take a normal state ω of $\mathcal{M}$ which is α-invariant and faifhful (i.e. such that $\varphi(A^*A) = 0$, $A \in \mathcal{M}$, implies $A = 0$: this condition is a natural one in the theory of Von Neumann algebras, although not physically cogent). Keeping the same stability requirements as above in Assumption (iii) and replacing Assumption (ii) by the requirement of "ergodicity of α " (= no α-invariant elements in $\mathcal{M}$ but the multiples of unity), we propose to establish the K.M.S. condition (3) for ω. Since, now, ω is assumed faithful, we know from the Tomita-Takesaki theory that ω generates a modular automorphism group σ for which it is K.M.S. at temperature 1 . Our game will therefore consist in proving the identity of α and σ up to scale factor β . The strategy is the following : we first note that α and σ commute, due to the fact that ω is α-invariant. Now the faithful and both α-invariant and σ-invariant ω generates a faithful representation of $\mathcal{M}$ in which α and σ are respectively implemented by unitary representation U and V ,whilst the two-parameter group $(t,s) \in R^2 \to \alpha_t\sigma_s$ is implemented by UV . Further, by a mechanism analogous to that which gave rise to the Proposition above Sp(U) , Sp(V) and Sp(UV) will all be groups, the latter a subgroup of R^2 whose projections on the x- and y-axes respectively coincide with Sp(U) and Sp(V) . Because of the assumed ergodicity of α , Sp(U) covers the whole reals. Sp(V) , on the other hand, can either be $\{ 0 \}$,or $\{ n\lambda; n \in \mathbb{Z} \}$ or $\hat{R}$. The first of these three cases trivially gives rise to stability (it corresponds to the temperature ∞ case in physics). The two others leave us with the three following possibilities for

1) an array of horizontal lines $y = n\lambda$, $n \in \mathbb{Z}$
2) the whole R^2-plane
3) a straight line of slope β through the origin.

We want to eliminate the two first cases and keep case 3) which leads to the desired proportionality of H and K , the infinitesimal

generators of respectively U, V ($U_t = e^{iHt}$, $V_s = e^{iKs}$).
Here is a sketch of the way in which this can be done :
if we rewrite (15) in the G.N.S. construction from ω introducing the modular operator $\Delta = e^K$ and the modular conjugation J of σ and using the fact that $h = h^*$ implies $JhJ = \Delta^{1/2} h$ we obtain the condition (Ω denotes the cyclic vector obtained from ω) :

$$(30) \qquad \omega_1^{(h)} (\alpha_T(A) - \alpha_S(A)) =$$

$$(i \int_S^T e^{-iHt} (I - \Delta^{1/2}) h \Omega \mid (I + \Delta^{1/2}) A \Omega)$$

$$, \ A = \pi(A) \ , \ h = \pi(h) \in \pi(\mathcal{M})$$

Now the limit $S \to -\infty$ (or $S \to -\infty$, $T \to +\infty$) will cause $\int_S^T e^{-iHt}$ to become, say, something like a principal value of $1/H$, while $I - \Delta^{-1/2} = I - e^{-1/2\ K}$ behaves like $\sim K$ in the spectral regions where K is small. Thus, roughly, the limit $S \to -\infty$ in () will make sense iff K/H is meaningful, a circumstance realized in case 3), but not in cases 1) and 2) for which there are regions of the UV -spectrum where $H = 0$ whilst K is finite. For the rigourization of this bold argument, it seems that we need a spectral concentration theorem believed to be true, but not yet formally proven, by our friends in the theory of Von Neumann Algebras.

So please allow a rugged, but pious physicist to end his talk with a prayer for the progress of Harmonic Analysis of Non-Commutative Systems !

Centre National de la Recherche Scientifique
Centre de Physique Théorique
31, Chemin Joseph Aiguier
13274 MARSEILLE Cedex 2

REFERENCES

[1] ARAKI,H.:
Publ. RIMS. Kyoto University 9, N° 1 (1973).

[2] ARAKI,H.:
Ann. Sci. Ecole Norm. Sup. 6, N° 1 (1973).

[3] ARVESON,W.:
On Groups of Automorphisms of Operators Algebras.
Preprint.

[4] BORCHERS,H.J.:
Nachr. Akad. Wiss. Göttingen II 2, 1 (1973).

[5] CONNES,A.:
Ann. Sci. Ecole Norm. Sup. 6, 18 (1973).

[6] DOPLICHER,S., KADISON,R.V., KASTLER,D., ROBINSON,D.W.:
Commun. Math. Phys. 6, 101 (1967).

[7] DOPLICHER,S., KASTLER,D., ROBINSON,D.W.:
Commun. Math. Phys. 3, 1 (1966).

[8] HAAG,R., HUGENHOLTZ,N., WINNINK,M.:
Commun. Math. Phys. 5, 215 (1967).

[9] HAAG,R., KASTLER,D., TRYCH-POHLMEYER,E.B.:
Commun. Math. Phys. 38, 173-193 (1974).

[10] KASTLER,D., POOL,J.C.T., THUE POULSEN,E.:
Commun. Math. Phys. 12, 175 (1969).

[11] KUBO,R.:
J. Physic. Soc. Japan 12, 570 (1957).

[12] MARTIN,P.C., SCHWINGER,J.:
J. Phys. Rev. 115, 1342 (1959).

[13] ROBINSON,D.W.:
Commun. Math. Phys. 31, 171 (1973).

[14] RUELLE,D.:
Commun. Math. Phys. 3, 133 (1966).

[15] STÖRMER,E.:
Commun. Math. Phys. 28, 279 (1972).

Verma Modules and the Existence of Quasi-Invariant Differential Operators

Bertram Kostant*

Introduction. Let σ be a representation of a Lie algebra $\mathfrak{g}$ by first order smooth (C^∞) differential operators on a manifold M. If α is a differential operator on M which commutes with $\sigma(y)$ for all $y \in \mathfrak{g}$ then certainly the space S of all $f \in C^\infty(M)$ such that $\alpha f = 0$ is stable under the action of $\mathfrak{g}$. However, one can in fact weaken the assumption of commutativity and still retain the stability of the space of solutions under the action of $\mathfrak{g}$. We will say that a differential operator α is quasi-invariant with respect to σ if for each $y \in \mathfrak{g}$ there exists a function $h^y \in C^\infty(M)$ such that the commutator

$$[\sigma(y), \alpha] = h^y \alpha .$$

It is clear that S is again stable under the action of $\mathfrak{g}$. One notes therefore that the existence of a quasi-invariant differential operator anticipates the non-irreducibility of the representation σ on $C^\infty(M)$.

There are a number of instances in Lie theory where one encounters quasi-invariance as opposed to strict invariance. We will call attention to two such instances.

In the first instance if $\mathfrak{g}$ is the Lie algebra of the conformal group $SO(4,2)$ then it is well known (particularly to physicists) that this is a multiplier representation σ of $\mathfrak{g}$ on the Minkowski space M^*

* This paper is partially supported by Grant No. P28969 of the National Science Foundation.

such that the wave operator $\Box = (\frac{\partial}{\partial x})^2 + (\frac{\partial}{\partial y})^2 + (\frac{\partial}{\partial z})^2 - (\frac{\partial}{\partial t})^2$ is quasi-invariant. This fact was pointed out to me by I.E. Segal. It is this which accounts for the also well known fact that the solutions of the wave equation are conformally invariant.

The second instance arises from the work of Zhelobenko. See [4] and also an earlier paper on the finite-dimensional representation of the classical groups. Zhelobenko constructs the finite-dimensional representations as solutions of certain quasi-invariant differential operators.

Both examples above are special cases of the following. Let G be a semi-simple Lie group with Lie algebra $\mathfrak{g}$ and let P be a parabolic subgroup. Then one knows that there is a nilpotent subgroup $\overline{N} \subseteq G$ such that $\overline{N}P$ is open in G. The $\overline{N}$-orbit $Y(x)$ (the use of x will be made clear later) of the origin o in $Y = G/P$ is then open in Y and is diffeomorphic to a Euclidean space. By induction any character $\lambda : P \longrightarrow \mathbb{C}^*$ defines a multiplier representation σ_λ of $\mathfrak{g}$ on $C^\infty(Y(x))$. We raise the question as to whether there exist quasi-invariant differential operators on $Y(x)$.

If we assume in addition, as we shall, that α is non-vanishing then the question reduces quickly to considering the existence of quasi-invariant differential operators in the set Γ of all $\overline{N}$-invariant differential operators on $Y(x)$.

In order to state the main theorem we first observe that the set of all distributions on $Y(x)$ with support at the origin is in a natural way a Verma module $V_{-\lambda}$ for $\mathfrak{g}$ with highest weight $-\lambda$. For the definition and properties of Verma modules, see [3], e.g. Dixmier [1], Gelfand, Gelfand and Bernstein [2].

One crucial property of Verma modules and, in a certain sense, a characterizing property of Verma modules, is the existence of leading weight vectors. See § 4.3. In case the complexification $p_{\mathbb{C}}$ of the

Lie algebra p of P is a Borel subalgebra the principal results in the paper [2] of I. Gelfand, S. Gelfand and N. Bernstein is the determination of the leading weight vectors. For the case of a general p progress in the determination of the leading weight vector has been made by Lepowsky.

The main result in this paper is a statement which reduces the question of quasi-invariance in Γ to leading weight vectors in $V_{-\lambda}$. If S is the Dirac measure at the origin the theorem asserts that the map $\alpha \rightarrow \alpha^t \delta$ sets up a bijection between the set of all quasi-invariant differential operators in Γ and all leading weight vectors in $V_{-\lambda}$. Thus for example, if $V_{-\lambda}$ is irreducible there are no quasi-invariant differential operators. (One has that $V_{-\lambda}$ is irreducible for an open dense set of λ in the set of all characters on P). The result is applied here for the case of the conformal group.

1. The underlying compact manifold $G/P = Y$ and the open submanifold $Y(x) \subseteq Y$.

1.1. Let G be any connected Lie group whose Lie algebra $\mathfrak{g}$ is reductive. We recall that $\mathfrak{g}$ is said to be reductive if $\mathfrak{g}$ is completely reducible with respect to the adjoint representation (i.e. $\mathfrak{g}$ is semi-simple plus abelian).

An element $x \in \mathfrak{g}$ is called real semi-simple if ad x is diagonalizable with real eigenvalues. If x is real semi-simple then x defines a linear direct sum decomposition

$$\mathfrak{g} = \bar{\underline{n}}(x) + \mathfrak{g}^x + \underline{n}(x)$$

where $\underline{n}(x)$ (resp. $\bar{\underline{n}}(x)$) is the subspace of $\mathfrak{g}$ spanned by all eigenvectors of ad x belonging to positive (resp. negative) eigenvalues and $\mathfrak{g}^x$ (the centralizer of x) is the kernel of ad x. Using the relation that if $[x,y_i] = \lambda_i y_i$, $i = 1,2$, $\lambda_i \in \mathbb{R}$ then $[x,y_1 + y_2] = (\lambda_1 + \lambda_2)(y_1 + y_2)$ it follows easily that $\underline{n}(x)$, $\underline{\bar{n}}(x)$ and $\mathfrak{g}^x$ are Lie subalgebras. It also follows that $\mathfrak{g}^x$ normalizes $\underline{n}(x)$ and hence $\underline{p}(x) = \mathfrak{g}^x + \underline{n}(x)$ is a Lie subalgebra of $\mathfrak{g}$, and

(1.1.1) $$\mathfrak{g} = \underline{\bar{n}}(x) + \underline{p}(x)$$ is a linear direct sum.

A Lie subalgebra $\underline{p} \subseteq \mathfrak{g}$ is called parabolic in case $\underline{p} = \underline{p}(x)$ for some real semi-simple element $x \in \mathfrak{g}$.

Proposition 1.1. _Any parabolic subalgebra_ $\underline{p} \subseteq \mathfrak{g}$ _is equal to its own normalizer in_ $\mathfrak{g}$.

Proof. Writing $\underline{p} = \underline{p}(x)$ where x is real semi-simple one has $x \in \mathfrak{g}^x \subseteq \underline{p}$. But clearly $[x, \underline{\bar{n}}(x)] = \underline{\bar{n}}(x)$ so that by (1.1.1) $\underline{\bar{n}}(x)$ does not meet the normalizer of $\underline{p}$. But then by (1.1.1) $\underline{p}$ is its own normalizer. QED

1.2. A Lie subgroup $P \subseteq G$, not necessarily connected, is called parabolic if its Lie algebra is parabolic. Recall that a Lie subgroup P is closed in G (since it has a separable base) if and only if its identity component P_o is closed in G. As a corollary to Proposition 1.1 one has

Proposition 1.2.1. _Any parabolic subgroup_ $P \subseteq G$ _is necessarily closed in_ G.

Now let $x \in \mathfrak{g}$ be real semi-simple and let $\bar{N}(x)$ be the connected Lie subgroup of G corresponding to $\underline{\bar{n}}(x)$. Also let P be any parabolic subgroup (not necessarily connected) whose Lie algebra is $\underline{p}(x)$. The following proposition is well known and is readily proved using the adjoint representation of G on $\mathfrak{g}$. Write $\bar{N} = \bar{N}(x)$ and $\underline{\bar{n}} = \underline{\bar{n}}(x)$.

Proposition 1.2.2. _Let_ $m = \dim \underline{\bar{n}}$. (1) _The group_ $\bar{N}$ _is closed in_ G, _simply connected and in fact is diffeomorphic to_ $\mathbb{R}^m$.
(2) $\bar{N} \cap P = (e)$. (3) $\bar{N}P$ _is open in_ G.

1.3. Let P be a parabolic subgroup and P^c be the normalizer of the identity component P_o of P. Thus $P \subseteq P^c$ and in fact if $\underline{p}$ is the Lie algebra of P then

(1.3.1) $$P^c = \{g \in G \mid \mathrm{Ad}\, g\, (\underline{p}) \subseteq \underline{p}\}.$$

Now by Proposition 1.1. $P^c \subseteq G$ is also a parabolic subgroup whose Lie algebra is $\underline{p}$. We will call P^c the completion of P. Also if P is parabolic then P is called complete if $P = P^c$.

Remark 1.3. Note that by Proposition 1.1 the map $p \to P^c$ sets up a bijection between all parabolic subalgebras of $\underline{g}$ and all complete parabolic subgroups of G.

Now let P be any parabolic subgroup of G and let Y be the G-homogeneous space G/P and let

$$\varepsilon : G \longrightarrow Y$$

be the projection map so that $\varepsilon(g) = \tilde{g}$ where $\tilde{g} = gP$. If e is the identity of G let $o = \tilde{e} \in Y$ and for any $r \in Y$ and $g \in G$ let $g \cdot r \in Y$ be the transform of r by g.

Now if the Lie algebra of P is written $\underline{p}(x)$ where x is a real semi-simple element let $m = \dim \underline{\bar{n}}(x)$ and let $Y(x) = \varepsilon(\bar{N}(x))$. As a corollary of Proposition 1.2.2 one has

Proposition 1.3. (1) $\dim Y = m$.

(2) $Y(x)$ *is an open submanifold of* Y *and* $Y(x)$ *is diffeomorphic to* $\mathbb{R}^m$.

(3) $\bar{N}(x) \cdot o = Y(x)$.

Somewhat less trivial, but also well known is

Theorem 1.3. *Let the notation be as above. Then the open submanifold* $Y(x)$ *in* Y *is dense in* $Y(\overline{Y(x)} = Y)$ *if and only if the parabolic subgroup* P *is complete. Furthermore in such a case* Y *is a compact manifold.*

Proof. (Sketched). The statement that $Y(x)$ is dense in Y is clearly equivalent to the statement that $\overline{N}P$ is dense in G. If P is not complete and $g \in P^c - P$ then one easily has that $\overline{N}gP$ is an open subset of G disjoint from $\overline{N}P$ showing that $\overline{N}P$ is not dense. On the other hand if P is complete then P contains the center of G and hence Y is a homogeneous space for the semi-simple group $\mathrm{Ad}\, G$. But then the Iwasawa decomposition for Ad G implies that any maximal compact subgroup of Ad G maps surjectively onto Y proving that Y is compact. The density of $Y(x)$ in Y also then follows easily using the Bruhat decomposition for the complexification of Ad G. QED

2. The multiplier representation σ_λ.

2.1. We retain the notation above. Let $x \in \mathfrak{g}$ be real semi-simple. Let P be a parabolic subgroup whose Lie algebra is $\underline{p}(x)$. Let $\overline{N} = \overline{N}(x)$ and let $Y = G/P$.

Now let Λ be a group parameterizing the group of all homomorphisms (characters) $\chi : P \longrightarrow \mathbb{C}^*$. For each $\lambda \in \Lambda$ let χ_λ denote the corresponding homomorphisms. Also for any $a \in P$ let $a^\lambda = \chi_\lambda(a)$.

Now let $\lambda \in \Lambda$ and R_λ denote the space of all C^∞ functions h on the open subset $\overline{N}P$ of G such that $h(ga) = a^{-\lambda}\, h(g)$ for all $a \in P$, $g \in \overline{N}P$. It is then clear from Proposition 1.3 that one has a linear isomorphism

$$i_\lambda : C^\infty(\overline{N}) \longrightarrow R_\lambda$$

where if $f \in C^\infty(\overline{N})$, $i_\lambda f$ is defined by $i_\lambda f\,(ba) = f(b)\, a^{-\lambda}$ for all $b \in \overline{N}$, $a \in P$.

Now for any $y \in \mathfrak{g}$ let η^y be the vector field on $\overline{N}P$ defined so that if $\phi \in C^\infty(\overline{N}P)$, $b \in \overline{N}P$, then $(\eta^y\phi)(b) = \frac{d}{dt}\,\phi(\exp - ty \cdot b)\Big|_{t=0}$. This is well defined since $\exp - ty \cdot b \in \overline{N}P$ for t sufficiently small.

Since η^y is just the right invariant vector field on G restricted to $\bar{N}P$ one has

Proposition 2.1.1 (1) For $y,z \in \mathfrak{g}$ one has $[\eta^y, \eta^z] = \eta^{[y,z]}$.

(2) R_λ is stable under η^y for any $y \in \mathfrak{g}$.

It follows from Proposition 2.1.1 that one has a representation

(2.1.1) $$\beta_\lambda : \mathfrak{g} \longrightarrow \text{End } R_\lambda$$

of $\mathfrak{g}$ on R_λ for any $\lambda \in \Lambda$ by defining $\beta_\lambda(y) = \eta^y | R_\lambda$.

We wish to view the representation β_λ of $\mathfrak{g}$ on R_λ defined by Proposition 2.1.1 in another way. For any $y \in \mathfrak{g}$ let ξ^y be the vector field on $Y(x) \cong Y = G/P$ defined so that if $\phi \in C^\infty(Y(x))$ and $r \in Y(x)$ then $(\xi^y \phi)(r) = \frac{d}{dt} \phi(\exp - ty \cdot r)\Big|_{t=0}$. Again this is well defined since $\exp ty \cdot r \in Y(x)$ for t sufficiently small. Also $[\xi^y, \xi^z] = \xi^{[y,z]}$ for $y,z \in \mathfrak{g}$. Thus if $\text{Diff } Y(x)$ denotes the algebra of all differential operators on $Y(x)$ then

(2.1.2) $$\sigma_0 : \mathfrak{g} \longrightarrow \text{Diff } Y(x)$$

is a homomorphism where $\sigma_0(y) = \xi^y$.

Now let $\tau : C^\infty(Y(x)) \longrightarrow C^\infty(\bar{N})$ be the algebra isomorphism defined by putting $(\tau f)(b) = f(\tilde{b})$ for $f \in C^\infty(Y(x))$ and hence one has a linear isomorphism

$$\tau_\lambda : C^\infty(Y(x)) \longrightarrow R_\lambda$$

for any $\lambda \in \Lambda$ where $\tau_\lambda = i_\lambda \circ \tau$. But now observe that for $\lambda,\mu \in \Lambda$ one has

(2.1.3) $$R_\lambda R_\mu \subseteq R_{\lambda+\mu}.$$

In particular R_0 is an algebra and τ_0 is an algebra isomorphism. In fact the following is obvious.

Proposition 2.1.2. The representations σ_0 and β_0 of $\mathfrak{g}$ are equivalent and the map τ_0 defines the equivalence.

Now more generally for any $\lambda \in \Lambda$ let $\sigma_\lambda : \mathfrak{g} \longrightarrow \text{End } C^\infty(Y(x))$ be the representation of $\mathfrak{g}$ on $C^\infty(Y(x))$ defined so that if $y \in \mathfrak{g}$ and $f \in C^\infty(Y(x))$ then $\sigma_\lambda(y)f$ is given by the relation

$$\tau_\lambda(\sigma_\lambda(y)f) = \beta_\lambda(y)\tau_\lambda f \ . \tag{2.1.4}$$

Remark 2.1.1. By definition σ_λ is equivalent to β_λ and τ_λ defines the equivalence. Also by Proposition 2.1.2 this definition of σ_0 agrees with the previous one.

We now observe that σ_λ is a "multiplier" representation.

Theorem 2.1. _For any_ $y \in \mathfrak{g}$, $\lambda \in \Lambda$ _the operator_ $\sigma_\lambda(y) \in \text{End } C^\infty(Y(x))$ _is a differential operator of deg_ 1 _so that we may write_

$$\sigma_\lambda : \mathfrak{g} \longrightarrow \text{Diff } Y(x).$$

In fact one has a linear map

$$h_\lambda : \mathfrak{g} \longrightarrow C^\infty(X(x))$$

so that if 1 _denotes the identity operator on_ $C^\infty(Y(x))$ _and we regard_ Diff $Y(x)$ _as a left module for_ $C^\infty(Y(x))$ (_using function multiplication_), _then_

$$\sigma_\lambda(y) = \xi^y + h_\lambda^y \cdot 1 \tag{2.1.5}$$

where $y \in \mathfrak{g}$ _and we write_ $h_\lambda^y = h_\lambda(y)$.

Proof. It suffices to prove that for any $\psi \in C^\infty(Y(x))$ one has the commutation relation

$$[\sigma_\lambda(y), \psi \cdot 1] = \xi^y \psi \cdot 1. \tag{2.1.6}$$

(In fact this proves that $\sigma_\lambda(y) - \sigma_0(y)$ comuutes with multiplication operators and hence must be a multiplication operator). But now if $f \in C^\infty(Y(x))$ then $\tau_\lambda \sigma_\lambda(y)\, \psi f = \eta^y \tau_\lambda \psi\, f$. But $\tau_\lambda \psi f = (\tau_0 \psi)(\tau_\lambda\, f)$. But then since η^y is a derivation one has $\tau_\lambda \sigma_\lambda(y)\psi\, f = \eta^y(\tau_0\psi)(\tau_\lambda f) = (\eta^y(\tau_0\ \psi))\tau_\lambda f + (\tau_0\psi)\eta^y\tau_\lambda f$. But $\eta^y \tau_0 \psi = \tau_0(\xi^y\psi)$ and $\eta^y\tau_\lambda f = \tau_\lambda(\sigma_\lambda(y)f$ Thus $\tau_\lambda(\sigma_\lambda(y)\psi f) = (\tau_0(\xi^y\psi))(\tau_\lambda f) + (\tau_0\psi)(\tau_\lambda\sigma_\lambda(y)f) = \tau_\lambda((\xi^y\psi)f + \psi\sigma_\lambda(y)f)$.

That is

$$\sigma_\lambda(y)\psi f = (\xi^y\psi)f + \psi\sigma_\lambda(y)f.$$

But this is the same statement as $[\sigma_\lambda(y), \psi \cdot 1] = \xi^y\psi \cdot 1$. QED

Remark 2.1.2. Note that if $y \in \underline{\bar{n}}(x)$ then $h_\lambda^y = 0$ for any λ. That is, $\sigma_\lambda(y) = \sigma_0(y)$. This is clear since one easily has the relation $\tau_\lambda(\sigma_0(y)) = \beta_\lambda(y)\tau_\lambda$ for $y \in \underline{\bar{n}}(x)$.

2.2. We wish to give a formula for $h_\lambda^y \in C^\infty(Y(x))$ for any $y \in \mathfrak{g}$, $\lambda \in \Lambda$. For any $\lambda \in \Lambda$ let $\lambda_* \in \mathfrak{g}'$ (the dual of $\mathfrak{g}$) be the linear functional defined so that $\lambda_*|\underline{\bar{n}}(x) = 0$ and $\lambda_*|\underline{p}(x)$ is the differential of χ_λ. That is if $y \in \underline{p}(x)$ then

(2.2.1) $$\langle\lambda_*, y\rangle = \frac{d}{dt}\chi_\lambda(\exp ty)\Big|_{t=0}.$$

We can now evaluate h_λ^y at the origin $o \in Y(x)$.

Lemma 2.2.1. <u>For any</u> $y \in \mathfrak{g}$, $\lambda \in \Lambda$, <u>one has</u>

$$h_\lambda^y(o) = \langle\lambda_*, y\rangle.$$

<u>Proof</u>. If 1 denotes the identity function on $Y(x)$ then clearly from (2.1.5) one has $\sigma_\lambda(y)1 = h_\lambda^y$ and hence $(\sigma_\lambda(y)1)(o) = h_\lambda^y(o)$. But now if $1_\lambda = \tau_\lambda 1$ then $(\sigma_\lambda(y)1)(o) = (\eta^y 1_\lambda)(e) = \frac{d}{dt}1_\lambda(\exp - ty)\Big|_{t=0}$.

Now write $y = y_1 + y_2$ where $y_1 \in \underline{\bar{n}}(x)$ and $y_2 \in \underline{p}(x)$. But if t is sufficiently small $\exp - ty \in \bar{N}P$ and hence for small t there exist curves $a_i(t)$, $i = 1,2$, so that $a_1(t) \in \bar{N}$, $a_2(t) \in P$ and $\exp - ty = a_1(t)\, a_2(t)$. Furthermore the tangent vector $a_i'(o)$ to $a_i(t)$ at $t = 0$ is just $-y_i$. Thus $1_\lambda(\exp - ty) = 1_\lambda(a_1(t)\, a_2(t)) = \chi_{-\lambda}(a_2(t))$. But then the derivative at $t = 0$ is just $\langle\lambda_*, y_2\rangle = \langle\lambda_*, y\rangle$ proving the lemma. QED

We now show that $h_\lambda^y \in C^\infty(Y(x))$ or rather $\tau h_\lambda^y \in C^\infty(\bar{N})$ is just a representative function for the coadjoint representation of $\bar{N}$ on $\mathfrak{g}'$. We first observe that $\bar{N}$ operates on $C^\infty(Y(x))$ (resp. $\mathbb{C}^\infty(\bar{N}P)$)

according to a representation σ (resp. β) where if $a \in \overline{N}$, $f \in C^\infty(Y(x))$ (resp. $f \in C^\infty(\overline{N}P)$) and $b \in Y(x)$ (resp. $b \in \overline{N}P$) then $(\sigma(a)f)(b) = f(a^{-1}b)$ (resp. $f(a^{-1}b)$). We assert

Lemma 2.2.2. If $a \in \overline{N}$, $\lambda \in \Lambda$ and $y \in \mathfrak{g}$ then $\sigma(a^{-1})\sigma_\lambda(y)\sigma(a) = \sigma_\lambda(\mathrm{Ad}\ a^{-1}(y))$.

Proof. It is immediate that $\tau_\lambda\sigma(a^{-1})\sigma_\lambda(y)\sigma(a) = \beta(a^{-1})\beta_\lambda(y)\beta(a)\tau_\lambda$. But the definition of the adjoint representation $\beta(a^{-1})\beta_\lambda(y)\beta(a) = \beta_\lambda(\mathrm{Ad}\ a^{-1}(y))$. Hence $\sigma(a^{-1})\sigma_\lambda(y)\sigma(a) = \sigma_\lambda(\mathrm{Ad}\ a^{-1}y)$ by definition of σ_λ. QED

One can determine the multiplier h_λ^y.

Theorem 2.2. Let $y \in \mathfrak{g}$, $\lambda \in \Lambda$, then for any $a \in \overline{N}$, one has

$$h_\lambda^y(\tilde{a}) = \langle \lambda_*, \mathrm{Ad}\ a^{-1}(y) \rangle .$$

Proof. We recall $\sigma_\lambda(y)1 = h_\lambda^y$. But $h_\lambda^y(\tilde{a}) = h_\lambda^y(a\cdot o) = (\sigma(a)^{-1}h_\lambda^y)(o)$ But $\sigma(a)^{-1}h_\lambda^y = \sigma(a)^{-1}\sigma_\lambda(y)1 = (\sigma(a^{-1})\sigma_\lambda(y)\sigma(a))1$ since $\sigma(a)1 = 1$. But then by Lemma 2.2.2 one has $\sigma(a^{-1})h_\lambda^y = \sigma_\lambda(\mathrm{Ad}\ a^{-1}(y))1$. Hence $h_\lambda^y(\tilde{a}) = (\sigma_\lambda(\mathrm{Ad}\ a^{-1}(y))1)(e)$. But then by Lemma 2.2.1 one has $h_\lambda^y(\tilde{a}) = \langle \lambda_*, \mathrm{Ad}\ a^{-1}y \rangle$. QED

We have observed that $h_\lambda^y = 0$ if $y \in \overline{\mathfrak{n}}(x)$. We now note that h_λ^y is constant if $y \in \mathfrak{g}^x$.

Corollary 2.2.1. Let $y \in \mathfrak{g}^x$, $\lambda \in \Lambda$, then the multiplier h_λ^y is the constant function with constant value $\langle \lambda_*, y \rangle$. In particular this is true for $y = x$.

Proof. We have only to observe that (since y^* vanishes on $\overline{\mathfrak{n}}(x)$) $a^{-1}y - y \in \overline{\mathfrak{n}}(x)$ for all $a \in \overline{N}$. But this is clear since obviously $[\overline{\mathfrak{n}}(x), \mathfrak{g}^x] \subseteq \overline{\mathfrak{n}}(x)$. QED

Now by nilpotence there clearly exists a positive integer k such that

(2.2.2) $$(\text{ad}\,\underline{\bar{n}}(x))^k \underline{\mathfrak{g}} \subseteq \underline{\bar{n}}(x).$$

In particular if $\underline{\bar{n}}(x)$ is commutative then we can choose $k = 2$.

Corollary 2.2.2. *Let* k *satisfy* (2.2.2). *Then for any* $z_i \in \underline{\bar{n}}(x)$, $i = 1,2,\ldots,k$, $y \in \underline{\mathfrak{g}}$ *and* $\lambda \in \Lambda$ *one has*

$$\xi^{z_1}\ldots\xi^{z_k} h^y_\lambda = 0.$$

In particular if $\underline{\bar{n}}(x)$ *is commutative* (*so that* $Y(x)$ *has a natural linear structure*) *then for any* $y \in \underline{\mathfrak{g}}$, $\lambda \in \Lambda$, *the function* h^y_λ *is, at most, a polynomial of degree* 1 *on* $Y(x)$.

Proof. The first statement is an immediate consequence of Theorem 2.2 and the fact that λ vanishes on $\underline{\bar{n}}(x)$. The second follows from the fact that we can take $k = 2$. QED

3. Quasi-invariance.

3.1 Now let $\Gamma \subseteq \text{Diff}\,Y(x)$ denote the algebra of all differential operators α on $Y(x)$ which are invariant under the action of $\bar{N}$ on $Y(x)$. That is, all $\alpha \in \text{Diff}\,Y(x)$ which commute with $\sigma(a)$ for all $a \in \bar{N}$, or equivalently all such α which commute with $\sigma_0(y)(=\sigma_\lambda(y)$ for any $\lambda \in \Lambda)$ for all $y \in \underline{\bar{n}}$ (writing $\underline{\bar{n}}$ for $\underline{\bar{n}}(x)$). If $y \in \underline{\bar{n}}$ let ρ^y denote the vector field on $Y(x)$ defined so that if $f \in C^\infty(Y(x))$, $a \in \bar{N}$ then $(\rho^y f)(a\cdot o) = \frac{d}{dt} f(a \exp ty\cdot o)\Big|_{t=0}$. It is clear that $\rho^y \in \Gamma$ and hence one obtains a map $\gamma : \underline{\bar{n}} \longrightarrow \Gamma$ by putting $\gamma(y) = \rho^y$. Furthermore if $U(\underline{\bar{n}})$ is the universal enveloping algebra (over $\mathbb{C}$) of $\underline{\bar{n}}$ then, since clearly $[\rho^y, \rho^z] = \rho^{[y,z]}$, if $y,z \in \underline{\bar{n}}$, the map γ extends to a homomorphism

(3.1.1) $$\gamma : U(\underline{\bar{n}}) \longrightarrow \Gamma\ .$$

Furthermore since an element in Γ is determined by its value at any one point of $Y(x)$ and that value may be arbitrary one clearly has

Proposition 3.1. <u>The map</u> γ <u>is an isomorphism of algebras</u>.

3.2. Let $\lambda \in \Gamma$. A nowhere vanishing differential operator $\alpha \in \operatorname{Diff} Y(x)$ will be said to be quasi-invariant (with respect to λ) if for each $y \in \underline{\mathfrak{g}}$ there exists a function $k_\alpha^y \in C^\infty(Y(x))$ such that the commutator

$$(3.2.1) \qquad [\sigma_\lambda(y), \alpha] = k_\alpha^y \alpha .$$

It is clear that if α is quasi-invariant so is $\phi\alpha$ for any nowhere vanishing $\phi \in C^\infty(Y(x))$.

Remark 3.2. The definition of quasi-invariance of α here includes the condition that α should be nowhere vanishing. Such an assumption would be automatically satisfied if quasi-invariant was defined group theoretically rather than infinitesimally. For our purpose it is so assumed in order to reduce to the case of $\overline{N}$-invariance. See Proposition 3.2.

It follows easily that the map $y \longmapsto k_\alpha^y$ defines a 1-cocycle on $\underline{\mathfrak{g}}$ in that for any $y, z \in \underline{\mathfrak{g}}$ one has the identity

$$(3.2.2) \qquad \xi^y k_\alpha^z - \xi^z k_\alpha^y = k_\alpha^{[y,z]} .$$

Now let $y_1, \ldots, y_m$ be a basis of $\underline{\overline{n}}$. Write $\xi^{y_i} = \xi_i$. Then since $Y(x)$ is diffeomorphic to $\overline{N}$ there exist (corresponding to right invariant 1-forms on $\overline{N}$) 1-forms ω_i, $i = 1,2,\ldots,m$ such that $\langle \omega_i, \xi_j \rangle = \delta_{ij}$ on $Y(x)$. If $f \in C^\infty(Y(x))$ it is then clear that

$$df = \sum_{i=1}^{m} (\xi_i f)\omega_i .$$

Now let ω be the 1-form on $Y(x)$ defined by putting

$$\omega = \sum k_\alpha^i \omega_i$$

where $k_\alpha^i = k_\alpha^{y_i}$.

It then follows easily from (3.2.2) that $d\omega = 0$. But since $Y(x)$ is simply connected one has $\omega = df$ for some $f \in C^\infty(Y(x))$ and hence one has

(3.2.3)
$$\xi^y f = k_\alpha^y$$

for all $y \in \bar{\underline{n}}$. But then if we put $\alpha_o = e^{-f}\alpha$ it follows that $[\xi^y, \alpha_o] = 0$ for all $y \in \bar{\underline{n}}$ and hence $\alpha_o \in \Gamma$. Furthermore $\alpha = e^f \alpha_o$. Thus one is reduced to considering quasi-invariance in Γ. That is, we have

Proposition 3.2. _A differential operator on_ $Y(x)$ _is quasi-invariant with respect to_ σ_λ _if and only if it is of the form_ $\phi\alpha$ _where_ $\alpha \in \Gamma$ _is quasi-invariant and_ ϕ _is a non-vanishing function on_ $Y(x)$.

3.3. Now let $\alpha \in \Gamma$ and $y \in \underline{\mathfrak{g}}$. We consider the question of computing $[\sigma_\lambda(y), \alpha]$. For $y \in \bar{\underline{n}}$ this vanishes. We next consider the case where $y \in \underline{\mathfrak{g}}^x$.

Let

(3.3.1)
$$P^x = \{g \in P \mid \text{Ad}\, gx = x\}.$$

The following proposition is easy and its proof is omitted. Let $N \subseteq P$ be the Lie subgroup corresponding to $\underline{n} = \underline{n}(x)$.

Proposition 3.3.1. (1) $P^x \cap N = (e)$.

(2) $P^x N = P$

(3) _The Lie algebra of_ P^x _is_ $\underline{\mathfrak{g}}^x$.

(4) P^x _normalizes_ $\bar{\underline{n}}$ _and hence_ $\bar{N}$.

Now since P^x normalizes $\bar{\underline{n}}$ it operates on $U(\bar{\underline{n}})$ by the adjoint representation. Similarly $\underline{\mathfrak{g}}^x$ operates on $U(\bar{\underline{n}})$ by the adjoint representation where if $z \in \underline{\mathfrak{g}}^x$, $u \in U(\bar{\underline{n}})$, then $(\text{ad}\, z)u = [z,u] = zu - uz \in U(\bar{\underline{n}}) \subseteq U$. Here and throughout $U = U(\underline{\mathfrak{g}})$ denotes the universal enveloping algebra, over $\mathbb{C}$, of $\underline{\mathfrak{g}}$.

Now $Y(x)$ is stable under the action of P^x since for any $a \in P^x$, $b \in \overline{N}$ one has $a\cdot(b\cdot o) = (aba^{-1})\cdot a\cdot o$. But $a\cdot o = o$ and $aba^{-1} \in \overline{N}$. Hence $a\cdot(b\cdot o) \in P^x$. We can then define a representation $\sigma : P^x \longrightarrow \text{Aut } C^\infty(Y(x))$ by putting $(\sigma(a)f)(r) = f(a^{-1}\cdot r)$ for $a \in P^x$, $f \in C^\infty(Y(x))$ and $r \in Y(x)$.

Remark 3.3. Note σ here is just an exponential of the representation $\sigma_0 | \mathfrak{g}^x$.

Recall that every $\alpha \in \Gamma$ is of the form $\gamma(u)$ for some $u \in U(\underline{\overline{n}})$.

Lemma 3.3. Let $a \in P^x$, $z \in \mathfrak{g}^x$ and $u \in U(\underline{\overline{n}})$. Then

$$(1) \quad \sigma(a)\gamma(u)\sigma(a^{-1}) = \gamma(\text{Ad } a(u))$$

$$(2) \quad [\xi^z, \gamma(u)] = \gamma([z,u]) .$$

Proof. It is enough to take $u = y \in \underline{\overline{n}}$ so that $\gamma(u) = \rho^y$. But if $f \in C^\infty(Y(x))$, $b \in N$, then $(\sigma(a)\rho^y\sigma(a^{-1})f)(b\cdot o) = \rho^y(\sigma(a^{-1})f)(a^{-1}ba\cdot o)$ $\frac{d}{dt}(\sigma(a^{-1})f)(a^{-1}ba \exp ty\cdot o)\Big|_{t=0} = \frac{d}{dt}f(b \exp t \text{ Ad } a(y)\cdot o)\Big|_{t=0} =$ $(\rho^{\text{Ad } a(y)}f)(b)$ proving (1). Now (2) follows since $([\xi^z, (u)]f)(b) =$ $\frac{d}{dt}(\sigma(\exp tz)\gamma(u)\sigma(\exp - tz)f)(b)\Big|_{t=0}$. See Remark 3.3. QED

As a corollary one has

Proposition 3.3.2. Let $z \in \mathfrak{g}^x$, $\alpha \in \Gamma$, $\lambda \in \Lambda$. Then $[\sigma_\lambda, \alpha]$ is again in Γ. In fact if we write $\alpha = \gamma(u)$ for $u \in U(\underline{\overline{n}})$, then

$$[\sigma_\lambda(z), \gamma(u)] = \gamma[z,u].$$

Proof. This is immediate from Lemma 3.3 and Corollary 2.2.1 since this statement implies that $\sigma_\lambda(z)$ differs from ξ^z by a constant. QED

3.4. An element $u \in U(\underline{\overline{n}})$ will be said to be a weight vector with weight $\upsilon \in \nu$ if $[z,u] = \langle \nu_*, z\rangle u$ for all $z \in \mathfrak{g}^x$.

Remark 3.4.1. We note that if $0 \neq u \in U(\underline{\bar{n}})$ and $[z,u] = \ell(z)u$ for all $z \in \underline{\mathfrak{g}}^x$ then there exists $\nu \in \Lambda$ such that $\ell(z) = \langle \nu_*, z \rangle$ for all $z \in \underline{\mathfrak{g}}^x$ so that u is a weight vector with weight ν. To see this let $G_{\mathbb{C}}$ denote the complex adjoint group of $\underline{\mathfrak{g}}_{\mathbb{C}}$, the complexification of $\underline{\mathfrak{g}}$. Then the adjoint repesentation defines a homomorphism $\mathrm{Ad} : G \longrightarrow G_{\mathbb{C}}$. Furthermore if $G_{\mathbb{C}}^x$ is the centralizer of x in $G_{\mathbb{C}}$ then one has

$$\mathrm{Ad}\ P^x \subseteq G_{\mathbb{C}}^x .$$

On the other hand the Lie algebra of $G_{\mathbb{C}}^x$ is $\mathrm{ad}\ \underline{\mathfrak{g}}_{\mathbb{C}}^x$ where $\underline{\mathfrak{g}}_{\mathbb{C}}^x$ is the complexification of $\underline{\mathfrak{g}}^x$. It follows then that there is a character $g \longrightarrow g^\tau$ in $G_{\mathbb{C}}^x$ such that $gu = g^\tau u$ for all $g \in G_{\mathbb{C}}^x$ where we extend the adjoint action of $G_{\mathbb{C}}$ to operate on $U = U(\underline{\mathfrak{g}})$. But then $a \longmapsto (\mathrm{Ad}\ a)^\tau$ defines a character on P^x and by Proposition 3.3.1 it extends to a character $a \longrightarrow a^\nu$, $\nu \in \Lambda$, on P. Thus one has $(\mathrm{Ad}\ a)u = a^\nu u$ for all $a \in P^x$ and hence, upon differentiating, one has $[z,u] = \langle \nu_*, z \rangle u$.

We will say that a differential operator $\alpha \in \Lambda$ has weight $\nu \in \Lambda$ if:

$$[\sigma_\lambda(z), \alpha] = \langle \nu_*, z \rangle \alpha \tag{3.4.1}$$

for all $z \in \underline{\mathfrak{g}}^x$.

Remark 3.4.2. If $u \in U(\underline{\bar{n}})$ and $\gamma(u) = \alpha$, then note that, by Proposition 3.3, u has weight $\nu \in \Lambda$ if and only if $\alpha \in \Gamma$ has weight ν.

Proposition 3.4. <u>If</u> $\alpha \in \Gamma$ <u>is a quasi-invariant of</u> σ_λ <u>then</u> α <u>has weight</u> ν <u>for some</u> $\nu \in \Lambda$. <u>Furthermore, in such a case, for any</u> $z \in \underline{\mathfrak{g}}^x$ <u>the function</u> k_α^z (<u>in</u> (3.2.1)) <u>is constant with constant value</u> $\langle \nu_*, z \rangle$.

<u>Proof</u>. Write $\alpha = \gamma(u)$, $u \in U(\underline{\bar{n}})$. If α is quasi-invariant then by Proposition 3.3.2 one has $[z,u] = \ell(z)u$ for all $z \in \underline{\mathfrak{g}}^x$ where ℓ is a linear functional on $\underline{\mathfrak{g}}^x$. But then by Remark 3.4.1 one has

$[z,u] = \langle \nu_*, z\rangle u$ for some $\nu \in \Lambda$. But then u has weight ν by Proposition 3.3.2. Comparing (3.2.1) with (3.4.1) one also has $\langle \nu_*, z\rangle = k_\alpha^z$ for any $z \in \mathfrak{g}^x$. QED

3.5. We can now determine k_α^y for any $y \in \mathfrak{g}$.

Proposition 3.5. *Assume $\alpha \in \Gamma$ is a quasi-invariant of weight $\nu \in \Lambda$ for the representation σ_λ. Then for any $y \in \mathfrak{g}$ one has*

$$k_\alpha^y = h_\nu^y$$

so that k_α^y may be given by Theorem 2.2.

Proof. It is an immediate consequence of the fact that $\sigma^y = \xi^y + h^y$ is a representation that one has the cocycle relation

$$\xi^y h_\nu^z - \xi^z h_\nu^y = h_\nu^{[y,z]} \tag{3.5.1}$$

for any $y,z \in \mathfrak{g}$ and hence if we put $j^y = k_\alpha^y - h_\nu^y$ then by (3.2.2) one also has $\xi^y j^z - \xi^z j^y = j^{[y,z]}$ for all $y,z \in \mathfrak{g}$. But now k_α^y and h_ν^y are constant for $y \in \mathfrak{g}^x + \bar{\underline{n}}(x) = \bar{p}(x)$ by Remark 2.1.2, Corollary 2.2.1 and Proposition 3.4. Thus if $y \in \bar{p}(x)$ and $z \in \mathfrak{g}$ is arbitrary one has

$$\xi^y j^z = j^{[y,z]}. \tag{3.5.2}$$

But now if we put $y = x$ then since ad x is non-singular on $\bar{\underline{n}}(x) + \underline{n}(x)$ and since $\xi_o^x = 0$ it follows from (3.5.2) that $j^z(o) = 0$ for $z \in \bar{\underline{n}}(x) + \underline{n}(x)$. But also $j^z(o) = 0$ for $z \in \mathfrak{g}^x$ since $k_\alpha^z(o) = \langle \nu_*, z\rangle = h^z(o)$ by Proposition 3.4 and Lemma 2.2.1. Thus $j^z(o) = 0$ for all $z \in \mathfrak{g}$. Finally $j^z = 0$ for all $z \in \bar{\underline{n}}(x)$ since $h_\nu^z = 0$ by Remark 2.1.2 and $k_\alpha^z = 0$ since $\alpha \in \Gamma$. We now prove that $j^z = 0$ for all $z \in \mathfrak{g}$. Let $\mathfrak{g}_k = \{y \in \mathfrak{g} \mid (\text{ad}\, \bar{\underline{n}}(x))^k y = 0\}$ so that $\mathfrak{g}_k \subseteq \mathfrak{g}_{k+1}$ and and $\mathfrak{g} = \mathfrak{g}_k$ for some k. Also $\mathfrak{g}_0 = 0$. One inducts on k. If $j^z = 0$ for all $z \in \mathfrak{g}_k$ then it follows that j^z is constant for all $z \in \mathfrak{g}_{k+1}$. Indeed this is clear since $\xi^y j^z = 0$ by (3.5.2) for any $y \in \bar{\underline{n}}(x)$. But $j^z(o) = 0$. Hence $j^z = 0$ for any $z \in \mathfrak{g}$. QED

Summarizing, if α is quasi-invariant of weight $\nu \in \Gamma$ for σ_λ then for any $y \in \mathfrak{g}$ one has

$$(3.5.3) \qquad [\sigma_\lambda(y), \alpha] = h_\nu^y \alpha .$$

Remark 3.5. We note that λ enters into (3.5.3) only in so far as α is quasi-invariant for σ_λ . The function h_ν^y depends only on the weight ν of α .

4. Verma modules.

4.1. Now σ_λ represents $\mathfrak{g}$ on $C^\infty(Y(x))$. But a differential operator operates on all distributions. Using Theorem 2.1 we may therefore extend σ_λ so it is a representation of $\mathfrak{g}$ on $\mathcal{D}'(Y(x))$, the space of all distributions on $Y(x)$, by putting $\sigma_\lambda(y)\phi = \xi^y \phi + h_\lambda^y \phi$ for any $\phi \in \mathcal{D}'(Y(x))$. Furthermore since σ_λ is a representation we may extend its domain to the universal enveloping algebra U of $\mathfrak{g}$. Hence $\mathcal{D}'(Y(x))$ is a U-module with respect to σ_λ . We note that if $f \in C^\infty(Y(x))$ and $\phi \in \mathcal{D}'(Y(x))$ has compact support then for $y \in \mathfrak{g}$.

$$(4.1.1) \qquad \langle \sigma_\lambda(y)\phi, f\rangle + \langle \phi, \sigma_{-\lambda}(y)f\rangle = 0.$$

This is clear since, using Theorem 2.2,

$$(4.1.2) \qquad h_{-\lambda}^y = -h_\lambda^y .$$

Now let V be the space of all distributions on $Y(x)$ with support at the origin. Then V is clearly stable under the action of σ_λ defining on V the structure of a U-module. We denote V as a U-module with respect to σ_λ by V_λ . Let $\delta \in V_\lambda$ be the Dirac measure at the origin.

Proposition 4.1. _For any_ $y \in \underline{p}$ _one has_

$$\sigma_\lambda(y)\delta = \langle \lambda_*, y\rangle \delta.$$

Proof. Since $\xi_o^y = 0$ for $y \in \underline{p}(x)$ one has $\sigma_\lambda(y) = h_\lambda^y(o)\delta$. But $h_\lambda^y(o) = \langle \lambda_*, y\rangle$ by Lemma 2.2.1. QED

4.2. Now if $y \in \underline{\bar{n}}(x)$ then $\sigma_\lambda(y) = \xi^y$. But since the map $y \mapsto (\xi^y)_o$ defines an isomorphism of $\underline{n}(\bar{x})$ with the tangent space $T_o(Y(x))$ of $Y(x)$ at the origin o if follows that $u \mapsto (\sigma_\lambda(y))_o$ defines an isomorphism of $U(\underline{\bar{n}})$ with the set of all differential operators at o. One therefore has

Proposition 4.2.1. V_λ _is a cyclic_ $U(\underline{\bar{n}})$ _module. In fact the map_

$$\pi : U(\underline{\bar{n}}) \longrightarrow V_\lambda$$

given by putting $\pi(u) = \sigma_\lambda(u)\delta$ _is a linear isomorphism (independent of_ λ).

Now let I_λ be the left ideal in U generated by $y - \langle\lambda_*, y\rangle$ for $y \in \underline{p}$. It follows from Proposition 4.1 that I_λ lies in the annihilator of δ. In fact if $\pi_\lambda : U \longrightarrow V_\lambda$ is the surjection defined by $\pi_\lambda(w) = \sigma_\lambda(u)\delta$ then one has the relation

$$\sigma_\lambda(u)\pi_\lambda(w) = \pi_\lambda(uw) \tag{4.2.1}$$

for all $u, w \in U$ and also

Proposition 4.2.2. _One has an exaxt sequence_

$$0 \longrightarrow I_\lambda \longrightarrow U \xrightarrow{\pi_\lambda} V_\lambda \longrightarrow 0$$

so that V_λ _as a_ U-_module is isomorphic to the quotient module_ U/I_λ.

Proof. This follows from Proposition (4.2.1) and the fact that

$$U = U(\underline{\bar{n}}) \oplus I_\lambda \tag{4.2.2}$$

as linear spaces. One proves (4.2.2) by noting that if I_λ^o is the ideal in $U(\underline{p})$ generated by $y - \langle\lambda_*, y\rangle$ for $y \in \underline{p}$ then $U(\underline{p}) = I_\lambda^o \oplus C\cdot 1$. But $U = U(\underline{\bar{n}}) \otimes U(\underline{p})$ as linear spaces by the Birkhoff-Witt theorem. One then has $I_\lambda = U(\underline{\bar{n}}) \otimes I_\lambda^o$ and hence (4.2.2). QED

4.3. If $\underline{p}$ is a solvable Lie algebra then the module U/I_λ is called a Verma module. Verma modules have been extensively studied. See e.g. [1] and [2]. For the case of an arbitrary parabolic Lie algebra we continue to call such modules Verma modules. In this generality they have been studied by Lepowsky. Thus V_λ is a Verma module. One knows that for "almost all" λ, V_λ is irreducible. In general, if we let $V_\lambda^{\bar{n}} = \{v \in V_\lambda \mid \sigma_\lambda(y)v = 0 \text{ for all } y \in \bar{\underline{n}}\}$ then one knows that $V_\lambda^{\bar{n}}$ is finite-dimensional. Clearly $\mathbb{C}\,\delta \subseteq V_\lambda^{\bar{n}}$. It follows easily that V_λ is U-irreducible if and only if $\mathbb{C}\delta = V_\lambda^{\bar{n}}$. In general a vector $0 \neq v \in V_\lambda$ will be called a leading weight vector with weight $\nu \in \Lambda$ if $\sigma_\lambda(y)v = \langle \nu_*, y\rangle v$ for all $y \in \underline{p}$. Note that δ, called the highest weight vector, is a leading weight vector of weight λ. Clearly any leading weight vector is in $V_\lambda^{\bar{n}}$. In particular, if V_λ is irreducible then, up to a scalar, δ is the only leading weight vector.

Now by the Birkhoff-Witt theorem one has $U = U(\bar{\underline{p}}) \otimes U(\underline{n})$ as linear spaces. But $U(\underline{n}) = U(\underline{n})\cdot n \oplus \mathbb{C}\cdot 1$ and hence one has a direct sum

$$U = U(\bar{\underline{p}}) \oplus U\cdot\underline{n} \tag{4.3.1}$$

where $U\cdot\underline{n}$ is the left ideal in U generated by $\underline{n}$. Thus given any $u \in U$ there exists a unique element $q_u \in U(\bar{\underline{p}})$ such that $u - q_u \in U(\underline{n})$. But now $U(\bar{\underline{p}}) = U(\bar{\underline{n}}) \otimes U(\mathfrak{g}^x)$.

Lemma 4.3.1. <u>Let</u> $y \in \mathfrak{g}$ <u>and let</u> $u \in U(\bar{\underline{n}})$ <u>be arbitrary, then</u>

$$q_{yu} \in U(\bar{\underline{n}})\,\mathfrak{g}^x + U(\bar{\underline{n}}).$$

<u>Proof</u>. If $U(\bar{\underline{n}})_k$ is the subspace spanned by all products $w_1 \dots w_j$, $w_i \in \bar{\underline{n}}$ where $j \leq k$ we will prove the lemma by induction on k and assume the lemma is true if $u \in U(\bar{\underline{n}})_k$. We may assume that $u \in U(\bar{\underline{n}})_{k+1}$ and in fact assume $u = w_1 \dots w_{k+1}$, $w_i \in \bar{\underline{n}}$. Now if $y \in \bar{\underline{n}}$ the result is obvious. If $y \in \mathfrak{g}^x$ the proof is also trivial since $yu = [y,u] + uy$. But $uy \in U(\bar{\underline{n}})\,\mathfrak{g}^x$ and $[y,u] \in U(\bar{\underline{n}})$.

Thus if $y \in \bar{\underline{p}}$

(4.3.2) $$q_{yu} = yu \in U(\bar{\underline{n}})\,\mathfrak{g}^x \oplus U(\bar{\underline{n}}).$$

Now if $y \in \underline{n}$ then $yu - [y,u] = uy \in U\cdot\underline{n}$ so that $q_{yu} = q_{[y,u]}$. But $[y,u] = \sum_{j=1}^{k+1} w_1 \ldots w_{j-1}[y,w_j] \ldots w_{k+1}$. But if $v_j = [y,w_j]\ldots w_{k+1}$ then $q_{yu} = \sum_j w_1 \ldots w_{j-1}\, q_{v_j}$ and $q_{v_j} \in U(\bar{\underline{n}})\,\mathfrak{g}^x \oplus U(\bar{\underline{n}})$ by induction. QED

Now if $w \in U$ and $q_w \in U(\bar{\underline{n}})\,\mathfrak{g}^x \oplus U(\bar{\underline{n}})$ let $r_w \in U(\bar{\underline{n}})\,\mathfrak{g}^x$ and $s_w \in U(\bar{\underline{n}})$ be the components of q_w so that $q_w = r_w + s_w$. The significance of this decomposition will be related to the decomposition $\sigma_\lambda(y) = \xi^x + h^y_\lambda$.

One has

Lemma 4.3.2. <u>Let</u> $y \in \mathfrak{g}$ <u>and</u> $u \in U(\bar{\underline{n}})$ <u>then</u>

(4.3.3) $$\xi^y\pi(u) = \pi(s_{yu}).$$

<u>Proof</u>. Since $h^z_\lambda = 0$ for all $z \in \bar{\underline{n}}$ one has

(4.3.4) $$\pi_\lambda | U(\bar{\underline{n}}) = \pi.$$

But for any y one has $\sigma_0(y) = \xi^y$. Thus we must show $\sigma_0(y)\pi_0(u) = \pi_0(s_{yu})$. But $\mathfrak{g}^x \in I_0$. Thus $\pi_0(r_{yu}) = 0$. Hence we must show $\sigma_0(y)\pi_0(u) = \pi_0(q_{yu})$. But $\pi_0(q_{yu}) = \pi_0(yu)$ and $\sigma_0(y)\pi_0(u) = \pi_0(yu)$ by (4.2.1). QED

Now since $U(\bar{\underline{n}})\,\mathfrak{g}^x = U(\underline{n}) \otimes \mathfrak{g}^x$ one has a linear mapping $U(\bar{\underline{n}})\,\mathfrak{g}^x \to U(\bar{\underline{n}})$, $r \mapsto r(\lambda)$, where if $r = uz$, $u \in U(\bar{\underline{n}})$, $z \in \mathfrak{g}^x$ then $r(\lambda) = \langle \lambda_*, z\rangle u$.

It is clear that

(4.3.5) $$r - r(\lambda) \in I_\lambda.$$

Lemma 4.3.3. <u>Let</u> $y \in \mathfrak{g}$ <u>and</u> $u \in U(\bar{\underline{n}})$ <u>then</u>

$$h^y_\lambda \pi(u) = \pi(r_{yu}(\lambda)).$$

Proof. One has $\sigma_\lambda(y)\pi_\lambda(u) = \pi_\lambda(q_{yu})$ since $\pi_\lambda(U\cdot n) = 0$. But from (4.3.2) and the subtraction of (4.3.3) from (4.2.1) one has $h^y\pi_\lambda(u) = \pi_\lambda(r_{yu})$. But $\pi_\lambda(r_{yu}) = \pi_\lambda(r_{yu}(\lambda)) = \pi(r_{yu}(\lambda))$ by (4.3.5.) QED

4.4. Now let (w,y) denote the Killing form B on $\mathfrak{g}$. Then one knows that $\underline{n}$ and $\underline{\bar{n}}$ are non-singularly paired with respect to B. Now let $\underline{n}_1 \subseteq \underline{n}$ be the orthogonal complement of $[\underline{\bar{n}},\underline{\bar{n}}]$ in $\underline{n}$. One knows that $\underline{n}_1$ (the "simple" part of $\underline{n}$) generates $\underline{n}$. It also has the property that

$$[\underline{n}_1,\underline{\bar{n}}] \subseteq \mathfrak{g}^x + \underline{\bar{n}} = \underline{\bar{p}}. \tag{4.4.1}$$

To prove (4.4.1) it suffices, since $\underline{\bar{p}}$ is the B-orthogonal complement of $\underline{\bar{n}}$ in $\mathfrak{g}$ to prove that $([\underline{n}_1,\underline{\bar{n}}],\underline{\bar{n}}) = 0$. But this is clear from the invariance of B since $(\underline{n}_1,[\underline{\bar{n}},\underline{\bar{n}}]) = 0$.

We have already observed (see (4.3.2)) that if $u \in U(\underline{\bar{n}})$ and $y \in \underline{\bar{p}}$ then $yu = q_{yu}$. We now observe

Lemma 4.4.1. <u>Let</u> $y \in \underline{\bar{n}}_1$ <u>and let</u> $u \in U(\underline{\bar{n}})$ <u>be arbitrary. Then</u>

$$q_{yu} = [y,u].$$

Proof. We first note that $q_{yu} = q_{[y,u]}$ since $y \in \underline{n}$. One proves that $q_{[y,u]} = [y,u]$ by induction on k as in the proof of Lemma 4.3.1. Assume $u = wu_1$ where $u_1 \in U(\underline{\bar{n}})_k$, $w \in \underline{\bar{n}}$, and the result is true for u_1. Now $[y,u] = [y,w]u_1 + w[y,u_1]$. But $w[y,u_1] \in U(\underline{\bar{n}})\mathfrak{g}^x \oplus U(\underline{\bar{n}})$ by induction. But also $[y,w]u_1 \in U(\underline{\bar{n}})\mathfrak{g}^x \oplus U(\underline{\bar{n}})$ by (4.4.1) and (4.3.2). Thus $[y,u] \in U(\underline{\bar{n}})\mathfrak{g}^x \oplus U(\underline{\bar{n}})$ and hence $q_{[y,u]} = [y,u]$. QED

Now if $q = r + s$ where $r \in U(\underline{\bar{n}})\mathfrak{g}^x$ and $s \in U(\underline{\bar{n}})$ put $q(\lambda) = r(\lambda) + s$.

The following is the crucial lemma.

Lemma 4.4.2. <u>Let</u> $y \in \underline{n}_1$, $w \in U(\underline{\bar{n}})$ <u>and let</u> $u \in U(\underline{\bar{n}})$ <u>be a weight vector of weight</u> (say) ν. <u>Then</u>

$$q_{ywu}(\lambda) - q_{ywu}(\lambda)u = r_{yw}(\nu)\ u + wq_{yu}(\lambda).$$

Proof. Now $[y,w] = q_{yw} = r_{yw} + s_{yw}$ by Lemma 4.4.1. Write $r_{yw} = \sum_i v_i z_i$ where $v_i \in U(\bar{\underline{n}})$ and $z_i \in \underline{\mathfrak{g}}^x$. Thus $r_{yw}u =$ $\sum v_i\ [z_i,u] + \sum v_i u z_i$. But $[z_i,u] = \langle \nu_*, z_i\rangle u$. Thus $r_{yw}u \in U(\bar{\underline{n}})\,\underline{\mathfrak{g}}^x \oplus U(\bar{\underline{n}})$ and one has

(4.4.2) $$r_{yw}(\lambda) = \sum_i \langle \lambda_*, z_i\rangle\ v_i$$

and

(4.4.3) $$(r_{yw}u)(\lambda) = (\sum_i \langle \lambda_* + \nu_*, z_i\rangle v_i)\,u.$$

But $[y,w]\ u = r_{yw}u + s_{yw}u$. Thus

(4.4.4) $$([y,w]u)(\lambda) = (\sum_i \langle \lambda_* + \nu_*, z_i\rangle v_i)u + s_{yw}u.$$

But $q_{ywu} = [y,wu] = [y,w]u + w\,[y,u]$ by Lemma 4.4.1.

Thus

$$q_{ywu}(\lambda) = ([y,w]u)(\lambda) + (w\,[y,u])(\lambda).$$

But $(w[y,u])(\lambda) = w([y,u](\lambda)) = wq_{yu}(\lambda)$ again by Lemma 4.4.1.

Hence recalling (4.4.4) one has

(4.4.5) $$q_{ywu}(\lambda) = (\sum_i \langle \lambda_* + \nu_*, z_i\rangle v_i)u + s_{yw}u + wq_{yu}(\lambda).$$

Now $q_{yw}(\lambda) = r_{yw}(\lambda) + s_{yw}$ and hence applying (4.4.2) one has

(4.4.6) $$q_{yw}(\lambda)u = (\sum_i \langle \lambda_*, z_i\rangle v_i)u + s_{yw}u.$$

Subtracting (4.4.6) from (4.4.5) one obtains

$$q_{ywu}(\lambda) - q_{yw}(\lambda)u = (\sum \langle \nu_*, z_i\rangle v_i)u + wq_{yu}(\lambda).$$

But then by (4.4.2) for $\lambda = \nu$ one has

$$q_{ywu}(\lambda) - q_{yw}(\lambda)u = (r_{yw}(\nu))u + wq_{yu}(\lambda). \qquad \text{QED}$$

4.5 If β is a differential operator on $Y(x)$ let β^t the transpose be the operator on V defined so that $\langle v, \beta f\rangle = \langle \beta^t v, f\rangle$ for all $v \in V$, $f \in C^\infty(Y(x))$. Note that by (4.1.1) one has

(4.5.1) $$(\sigma_\lambda(y))^t = -\sigma_{-\lambda}(y) \mid V.$$

For $\beta = \gamma(u) \in \Gamma$ one has

Lemma 4.5.1. _Let_ $w, u \in U(\underline{\bar{n}})$ _then_

$$(\gamma(u))^t \pi(w) = \pi(wu).$$

Proof. By Proposition 3.1 the map $u \mapsto \gamma(u)^t$ is an anti-homomorphism of $U(\underline{\bar{n}})$. Thus it suffices to prove the lemma for the case where $u = y \in \underline{\bar{n}}$. But $\gamma(y)$ is the vector field ρ^y and ρ^y commutes with $\sigma_0(U(\underline{\bar{n}}))$. Thus using (4.5.1) $\gamma(y)^t\pi(w) = (\rho^y)^t\pi(w) = (\rho^y)^t\sigma_0(w)\delta = (\rho^y)^t(\sigma_0(-w))^t\delta = \sigma_0(-w)^t(\rho^y)^t\delta = \sigma_0(w)(\rho^y)^t\delta$. But $(\rho^y)^t\delta = \xi^y\delta = \sigma_0(y)\delta$ (since for any $f \in C^\infty(Y(x))$, $(\rho^y f)(o) = -(\xi^y f)(o))$. Thus $\gamma(y)^t\pi(w) = \sigma_0(w)\,\sigma_0(y)\delta = \sigma_0(wy)\delta = \pi(wy)$. QED

As a corollary one has

Proposition 4.5. _For any_ $u \in U(\underline{\bar{n}})$ _one has_

$$\gamma(u)^t\delta = \pi(u).$$

In particular, the map $\Gamma \longrightarrow V$ _given by_ $\alpha \mapsto \alpha^t\delta$ _is a bijection._

Proof. The first statement is proved by putting $w = 1$ as in Lemma 4.5.1 and noting that $\pi(1) = \delta$. The second statement is then a consequence of Proposition 3.1 and Proposition 4.2.1. QED

4.6. Now V is a Verma module $V_{-\lambda}$ with respect to the restriction of $\sigma_{-\lambda}$ to V. The main commutation we need is

Lemma 4.6. _Let_ $\alpha \in \Gamma$ _have weight_ $\nu \in \Gamma$. _Then for any_ $y \in \underline{n}_1$ _there exists_ $\alpha_0 \in \Gamma$ _such that_

(4.6.1) $$[\sigma_\lambda(y), \alpha] = h_\nu^y\alpha + \alpha_0$$

where in fact α_0 _is uniquely given by the relation_

(4.6.2) $$\sigma_{-\lambda}(y)\alpha^t\delta = \alpha_o^t\delta .$$

Proof. To prove the equality (4.6.1) it is enough to prove the equality at one point since the differential operators involved are analytic. Thus it suffices to prove the equality of the respective transpose operators on $V_{-\lambda}$. Now $(\sigma(y))^t = -\sigma_{-\lambda}(y)$ restricted to V_λ by (4.5.1). Thus $[\sigma_\lambda(y),\alpha]^t = [\sigma_{-\lambda}(y),\alpha^t]$ on V_λ. Let $w \in U(\underline{\bar{n}})$. Thus if we write $\alpha = \gamma(u)$, $u \in U(\underline{\bar{n}})$ then

(4.6.3) $$[\sigma_{-\lambda}(y),\alpha^t]\pi(w) = \sigma_{-\lambda}(y)\pi(wu) - \alpha^t\sigma_{-\lambda}(y)\pi(w)$$

by Lemma 4.5.1. But now $\sigma_\lambda(y) = \xi^y + h_\lambda^y$. Hence now, by Lemmas 4.3.2 and 4.3.3, for any $v \in U(\underline{\bar{n}})$

(4.6.4) $$\sigma_\lambda(y)\pi(v) = \pi(q_{yv}(\lambda)).$$

One thus has

$$\sigma_{-\lambda}(y)\pi(wu) = \pi(q_{ywu}(-\lambda))$$

and by Lemma 4.5.1

$$\alpha^t\sigma_{-\lambda}(y)\pi(w) = \pi(q_{yw}(-\nu)u).$$

But then (4.6.3) implies

$$[\sigma_{-\lambda}(y),\alpha^t]\pi(w) = (q_{ywu}(-\lambda) - q_{yw}(-\lambda|u)).$$

Now applying Lemma 4.4.2 and recalling Remark 3.4.2 one has $[\sigma_{-\lambda}(y),\alpha^t]\pi(w) = \pi(r_{yw}(\nu)u) + \pi(wq_{yu}(-\lambda))$. But by Lemmas 4.5.1 and 4.3.3 $\pi(r_{yw}(\nu)u) = \alpha^t\pi(r_{yw}(\nu)) = \alpha^t h_\nu^y\pi(w)$. But $\pi(wq_{yu}(-\lambda)) = \gamma(q_{yu}(-\lambda))^t\pi(w)$ by Lemma 4.5.1 and hence since $w \in U(\underline{\bar{n}})$ is arbitrary

$$[\sigma_{-\lambda}(y),\alpha^t] = (\alpha^t h_\nu^y + \gamma(q_{yu}(-\lambda))^t)$$

on V_λ. Taking transposes one has

$$[\sigma_\lambda(y), \alpha] = h_\nu^y\alpha + \alpha_o$$

where $\alpha_o^t\delta = \gamma(q_{yu}(-\lambda))^t\delta = \pi(q_{yu}(-\lambda))$ by Proposition 4.5. But $\pi(q_{yu}(-\lambda)) = \sigma_{-\lambda}(y)\pi(u)$ by (4.6.4). Thus $\alpha_o^t\delta = \sigma_{-\lambda}(y)\pi(u) = \sigma_{-\lambda}(y)\alpha^t\delta$

by Proposition 4.5. One notes also by Proposition 4.5 that this uniquely characterizes $\alpha_o \in \Gamma$. QED

4.7. We can now state our main theorem on the existence of quasi-invariant differential operators on the open set Y(x) on G/P.

Theorem 4.7. Let χ_λ be any character on P and σ_λ be the representation $\mathfrak{g}$ by differential operators on Y(x) given in § 2. Then an N-invariant (with respect to $\sigma_\lambda|\bar{N}$) differential operator α is a quasi-invariant (i.e. $\alpha \in \Gamma$) on Y(x) for σ_λ. (That is, $[\sigma_\lambda(y),\alpha] = k^y\alpha$ for some $k^y \in C^\infty(Y(x))$ for all $y \in \mathfrak{g}$) if and only if $\alpha^t\delta \in V_{-\lambda}$ is a leading weight vector for the representation $\sigma_{-\lambda}$ restricted to $V_{-\lambda}$. Here $V_{-\lambda}$ is the set of all distributions with support at the origin o of Y(x) and is a Verma module with highest weight $-\lambda$ with respect to the restriction of $\sigma_{-\lambda}$ to $V_{-\lambda}$. Also δ is the Dirac measure at o. The map $\alpha \mapsto \alpha^t\delta$ sets up a bijection between all quasi-invariant $\alpha \in \Gamma$ and all leading weight vectors in the Verma module $V_{-\lambda}$. Moreover if $\alpha \in \Gamma$ is quasi-invariant of weight ν then $k^y = h^y_\nu$ is given by Theorem 2.2.

Proof. If α is quasi-invariant then α has weight ν since $\nu \in \Lambda$ by Proposition 3.4. But then for any $y \in \mathfrak{g}$ one has $[\sigma_\lambda(y),\alpha] = h^y_\nu\,\alpha$ by (3.5.3) and hence taking transposes one has $[\sigma_{-\lambda}(y),\alpha^t] = \alpha^t h^y_\nu$ on $V_{-\lambda}$. But then one has $\sigma_{-\lambda}(y)\alpha^t\delta = \alpha^t h^y_\nu\delta + \alpha^t\sigma_{-\lambda}(y)\delta$. But $\sigma_{-\lambda}(y)\alpha^t = \xi^y + h^y_{-\lambda}$ so that

$$\sigma_{-\lambda}(y)\alpha^t\delta = \alpha^t(\xi^y + h^y_{\nu-\lambda})\delta = \alpha^t\sigma_{\nu-\lambda}(y)\delta. \tag{4.7.1}$$

But if $y \in \underline{n}$ then $\xi^y_o = 0$ and hence

$$\sigma_{-\lambda}(y)\alpha^t\delta = \alpha^t h^y_{\nu-\lambda}\delta = \langle \nu_* - \lambda_*, y\rangle\, \alpha^t\delta \tag{4.7.2}$$

since $h^y_{\nu-\lambda}(o) = \langle \nu_* - \lambda_*, y\rangle$ by Lemma 2.2.1. Thus $\alpha^t\delta$ is a leading weight vector with weight $\nu - \lambda$.

Conversely if $\alpha^t\delta$ is a leading weight vector in $V_{-\lambda}$ then one has the relation (4.7.3) for some $\nu \in \Lambda$. But since $\sigma_{-\lambda}(y)\delta = -\langle\lambda_*, y\rangle\delta$ for all $y \in \underline{p}$ one has the relation

$$(4.7.3) \qquad [\sigma_{-\lambda}(y), \alpha^t]\delta = \langle\nu_*, y\rangle\alpha^t\delta$$

for all such y. But if $y \in \underline{g}^x$ then $[\sigma_\lambda(y), \alpha] \in \Gamma$. To prove this it suffices to show that $[\sigma_\lambda(y), \alpha]$ commutes with $\sigma_0(\bar{\underline{n}}) = \sigma(\bar{\underline{n}})$. But this is clear since $\underline{g}^x$ normalizes $\bar{\underline{n}}$. Thus if $u \in U(\bar{\underline{n}})$ and we apply $\sigma_{-\lambda}(u)$ to both sides in 4.7.3, it follows from the commutativity that $[\sigma_{-\lambda}(y), \alpha^t]\,\pi(u) = \langle\nu_*, y\rangle\alpha^t\pi(u)$. Hence $[\sigma_{-\lambda}(y), \alpha^t] = \langle\nu_*, y\rangle\alpha^t$ on $V_{-\lambda}$ by Proposition 4.2.1. Thus taking transposes one has the relation $[\sigma_\lambda(y), \alpha] = \langle\nu_*, y\rangle\alpha$ for all $y \in \underline{g}^x$. Thus α has weight ν and we can apply Lemma 4.6. But then $\alpha_0 = 0$ since $\alpha^t\delta$ is a leading weight vector. Hence $[\sigma_\lambda(y), \alpha] = h^y\alpha$ for all $y \in \underline{n}$. But then since $\underline{n}_1$, $\underline{g}^x$ and $\bar{\underline{n}}$ generate $\underline{g}$ one has $[\sigma_\lambda(y), \alpha] = f^y\alpha$ for all $y \in \underline{g}$ so that α is quasi-invariant.

The remaining statements follow from (3.5.3) and Proposition 4.5. QED

Remark 4.7. Note we have observed that if α is quasi-invariant of weight ν the corresponding leading weight vector $\alpha^t\delta$ in the Verma module $V_{-\lambda}$ has weight $\nu - \lambda$.

Corollary 4.7. *If the Verma module* $V_{-\lambda}$ *is irreducible there are no non-trivial quasi-invariant differential operators on* $Y(x)$ *for the representation* σ_λ *of* $\underline{g}$.

5. *The case of the conformal group.*

5.1. We wish to apply the main theorem to the case where G is the conformal group and u corresponds to the wave operator. That is $G = SO(4,2)$ so that $\dim \underline{g} = 15$. The element $x \in \underline{g}$ that we shall take (up to normalization) is sometimes called the scale. It is characterized by the property $\dim \mathfrak{g}^x = 7$ and $[\mathfrak{g}^x, \mathfrak{g}^x]$ is the Lie algebra of the homogeneous Lorentz group $(SO(3,1))$. In particular $\mathfrak{g}^x = Rx \oplus [\mathfrak{g}^x, \mathfrak{g}^x]$. The element x is normalized up to size by the condition that the eigenvalues of ad x are $0, \pm 1$. The parabolic subalgebra $\underline{p}(x) = \mathfrak{g}^x + \underline{n}(x)$ is 11-dimensional and $\underline{n}(x)$ is an abelian 4-dimensional subalgebra on which ad $x = +1$. Then $\underline{\bar{n}}(x) = \underline{\bar{n}}$ is also 4-dimensional and abelian and one knows that there is a basis z_0, z_1, z_2, z_3 of $\underline{\bar{n}}$ such that the element $u = -z_0^2+z_1^2+z_2^2+z_3^2 \in U(\underline{\bar{n}})$ is invariant under Ad $(P^x)'$ where $(P^x)'$, the commutator subgroup of P^x, is the homogeneous Lorentz group. In fact u is a weight vector for some $\nu \in \Lambda$ where $\langle \nu_*, x \rangle = -2$.

The compact 4-dimensional manifold $Y = G/P$ has been called the cosmos by Segal. It has also been considered by Penrose and others. The open submanifold $Y(x) \subseteq Y$ may be identified with Minkowski spaces M^4. The transitive group $\bar{N}$ operating on $Y(x)$ is the group of spaces of space-time translations. Also $\gamma(u) \in \mathrm{Diff}\, Y(x)$ is the wave operator.

We raise the question: for which $\lambda \in \Lambda$ is the wave operator $\gamma(u)$ quasi-invariant for the representation σ_λ of $\mathfrak{g}$ on $C^\infty(Y(x))$? For such a the solution of the wave equation $\gamma(u)f = 0$ will be stable under $\sigma_\lambda(\mathfrak{g})$.

Now by Theorem 4.7 a necessary and sufficient condition for this is that $\gamma(u)^t = \delta = \sigma_{-\lambda}(u)\delta$ should be a leading weight vector in the Verma module $V_{-\lambda}$. For this to be the case we need only that $\sigma_{-\lambda}(\underline{n})\sigma_{-\lambda}(u)\delta = 0$. But since $\sigma_{-\lambda}(\underline{n})\delta = 0$ we need only the condition that $\sigma_{-\lambda}[n,u]\delta = 0$.

Now if B is the Killing form on $\mathfrak{g}$ then using the pairing of $\underline{n}$ and $\underline{\bar{n}}$ with respect to B the element u defines a quadratic form on $\underline{n}$. But this defines a linear map $\underline{n} \longrightarrow \underline{\bar{n}}$, $w \longrightarrow \bar{w}$. It is then easy to prove that if J is the left ideal in $U(\underline{n})$ generated by $[p,p]$ one has

(5.1.1) $$[w,u] - \frac{w}{4}(1 + x) \in J$$

for all $w \in \underline{n}$.

But $J \subseteq I_{-\lambda}$, using the notation of § 4.2 and hence

(5.1.2) $$[w,u]\ (-\lambda) = \frac{w}{4}\ (1 - \langle \lambda_*, x \rangle).$$

Thus $\sigma_{-\lambda}(\underline{\bar{n}})\sigma_{-\lambda}(u)\delta = 0 \Longleftrightarrow \langle \lambda_*, x \rangle = 1$. But $\langle \lambda_*, x \rangle$ characterizes λ.
Hence we have proved

Theorem 5.1. <u>There exists a unique character</u> $\lambda \in \Lambda$ <u>such that the wave operator</u> $\gamma(u)$ <u>is quasi-invariant for</u> σ_λ . <u>The element</u> λ <u>is characterized by the condition that</u> $\langle \lambda_*, x \rangle = 1$ <u>and since</u> ad $x = 1$ <u>on</u> $\underline{n}$ <u>this</u> λ <u>is the root on</u> $\mathbb{R}x$ <u>defined by the root space</u> $\underline{n}$.

REFERENCES

1. J. Dixmier, <u>Algèbres Enveloppantes</u>, Gauthier-Villars, Paris, 1974.
2. I.N. Bernstein, I.M. Gelfand, and S.I. Gelfand, <u>Structure of representations generated by highest weight vectors</u>, Funct. Anal., i evo prilogenie, 5, 1971, 1-9.
3. D.N. Verma, <u>Structure of certain reduced representations of complex semi-simple Lie algebras</u>, Bull. Amer. Math. Soc., 74, 1968, 160-166, 628.
4. D. Zhelobenko, <u>The analysis of irreducibility in the class of elementary representations of a complex semi-simple Lie group</u>, Izv. Akad. Nau., SSSR, Ser. Mat., 1968, 108-133.

Massachusetts Institute of Technology
Department of Mathematics
CAMBRIDGE - Massachusetts 02139 / USA

Sur la racine carrée du noyau de Poisson dans les espaces symétriques

Noël LOHOUE

0. Soit G un groupe de Lie semi-simple non compact, connexe et de centre fini. Soit K un sous-groupe compact maximal de G. Soit B la frontière de Furstenberg de l'espace symétrique $X = G/K$.

1. Soit G un groupe de Lie semi-simple, non compact, connexe et de centre fini. Soient $G = KAN$ et $G = KA^+K$ des décompositions d'Iwasawa et Cartan respectivement. Notons $\mathfrak{a}^+$ la chambre de Weyl positive correspondante ; soit Σ^+ l'ensemble des racines positives restreintes, chacune intervenant autant de fois que l'exige sa multiplicité, et notons 2ρ leur somme. On sait que, dans la décomposition de Cartan, une mesure de Haar sur G est donnée par la formule

$$(1)\qquad \int_G \varphi(g)\,dg = \int_K \int_{\mathfrak{a}^+} \int_K \varphi(k_1.\exp H.k_2) \prod_{\alpha\in\Sigma^+} \mathrm{sh}(\alpha(H))\,dk_1\,dH\,dk_2\ ,$$

où $dk_1 = dk_2$ est la mesure de Haar normalisée sur K, et où dH est la mesure sur $\mathfrak{a}^+$ restriction d'une mesure de Lebesgue sur l'algèbre de Lie $\mathfrak{a}$ de A.

Sur l'espace symétrique $X = G/K$, soit dx la mesure-quotient de dg par

K. Si M est le centralisateur de A dans K, la frontière de Furstenberg de X est l'espace homogène $B = K/M$ isomorphe à G/MAN. Sur B, soit $db = dk_M$ la mesure-quotient de dk par M.

Le noyau de Poisson sur $X \times B$ est donné par la formule

$$P(g\ ,kM) = e^{-2\rho(H(g^{-1}k))} \quad , \quad (g\in G\ ,\ k\in K)\ ,$$

où $\exp\left[H(g^{-1}k)\right]$ est la composante dans A de $g^{-1}k$ pour la décomposition d'Iwasawa.

THEOREME 1. Soit $B = K/M$ la frontière de Furstenberg d'un espace symétrique $X = G/K$. Pour tous nombres réels $q > 2$, $s > 2$ et pour toutes fonctions $U \in L^s(B)$ et $V \in L^2(B)$, la fonction $F(g) = (\Pi(g)V, U)$ où Π est la représentation quasi-régulière de G sur $L^2(B)$ est de puissance qième intégrable.

Preuve du théorème 1. On suppose d'abord que $2 < q \leq s$, alors on voit que la norme de F dans $L^q(G)$ est dominée par $\|U\|_{L^s(B)} \|F'\|_{L^q(G)}$ où F' est la fonction $F'(g) = \langle \Pi(g)V, 1\rangle$.

L'inégalité de convexité pour les normes L^p permet de voir qu'il en est ainsi pour tout $q > 2$.

Pour terminer la preuve il suffit d'utiliser le résultat suivant (3) qui est un travail commun avec P. Eymard.

THEOREME 2. Soit $B = K/M$ la frontière de Furstenberg d'un espace symétrique $X = G/K$. Pour tout nombre réel $q > 2$, par la formule

(2) $$f \to F, \quad \text{où} \quad F(gK) = \int_B P^{1/2}(gK, b)\, f(b)\, db,$$

la racine carrée du noyau de Poisson applique continûment $L^2(B, db)$ dans $L^q(X, dx)$.

Remarques :

1) Pour $H \in \mathfrak{a}^+$ et $k \in K$, posons désormais :

(3) $$P_H(k) = P((\exp H)\,K,\ kM).$$

Alors, si $g = k_1 . \exp H . k_2$ est la décomposition de Cartan de g, on a

$$F(gK) = \int_{K/M} P^{1/2}((k_1 \exp H)K, kM)\, f(kM)\, dk_M = \int_{K/M} P^{1/2}((\exp H)\, K, k_1^{-1}kM)\, f(kM)\, dk_M ,$$

soit

$$F(gK) = (\sqrt{P_H} * \check{\check{f}})^{\vee}(k_1) ,$$

où le produit de convolution est calculé sur K, les fonctions sur B étant identifiées à des fonctions sur K constantes sur les classes modulo M. (On note $\check{\varphi}(k) = \varphi(k^{-1})$).

2) Si $\alpha \in \Sigma^+$ et $H \in \mathfrak{a}^+$, on a $\operatorname{sh}(\alpha(H)) \leq e^{\alpha(H)}$, donc

$$\prod_{\alpha \in \Sigma^+} \operatorname{sh}(\alpha(H)) \leq e^{\sum_\alpha \alpha(H)} = e^{2\rho(H)} .$$

Ces remarques et la formule (1) montrent que, pour établir le Théorème 2, il suffit d'établir la

PROPOSITION. <u>Pour tout</u> $q > 2$, <u>il existe une constante</u> C_q <u>telle que, pour toute</u> $f \in L^2(K/M)$, <u>on ait</u> :

(4)
$$\int_{\mathfrak{a}^+} \left\|\sqrt{P_H} * f\right\|_q^q e^{2\rho(H)}\, dH \leq C_q \left\|f\right\|_2^q .$$

La démonstration de cette Proposition consiste essentiellement à appliquer <u>in the right place</u> le théorème d'interpolation de Marcinkiewicz (cf [4], tome II, p. 112). Pour plus de clarté, nous la décomposerons en lemmes.

LEMME 1. <u>Soit</u> ν <u>un nombre réel tel que</u> $1 < \nu \leq 2$. <u>Alors il existe une constante</u> c_ν <u>telle que, pour tout</u> $H \in \mathfrak{a}^+$, <u>on ait</u> :

$$\left\|\sqrt{P_H}\right\|_\nu \leq c_\nu e^{(1-\frac{2}{\nu})\rho(H)} .$$

Démonstration. Posons $\lambda = i(\nu - 1)\rho$, donc $i\lambda - \rho = -\nu\rho$. D'après la formule intégrale de Harish-Chandra pour les fonctions sphériques, on a, en posant $a = \exp H$,

$$\|\sqrt{P_H}\|_\nu^\nu = \int_K e^{-\nu\rho(H(a^{-1}k))}\,dk = \int_K e^{(i\lambda-\rho)(H(a^{-1}k))}\,dk = \varphi_\lambda(a^{-1}) = \varphi_{-\lambda}(a).$$

Comme $1 < \nu \leq 2$, on sait (cf [2], pp. 279 à 291, et aussi [3], p. 390) que, pour tout $H \in \mathfrak{a}^+$, si l'on pose $a_t = \exp(tH)$, la fonction de variable réelle

$$t \to \varphi_{-\lambda}(a_t)\, e^{(i\lambda+\rho)(tH)}$$

tend, quand $t \to +\infty$, en croissant vers la constante $\underline{c}(-\lambda)$, où $\underline{c}$ est la fonction d'Harish-Chandra. Remarquant que $i\lambda + \rho = (2 - \nu)\rho$, on en déduit le lemme en posant $c_\nu = [\underline{c}(-\lambda)]^{1/\nu}$.

DEFINITION. Pour $H \in \mathfrak{a}^+$ et $f \in L^1(K/M)$, posons

$$Tf(H) = e^{(1+\frac{2}{q})\,\rho(H)} \| \sqrt{P_H} * f \|_q , \tag{5}$$

où le nombre réel $q > 2$ est désormais fixé.

LEMME 2. <u>Si</u> $\frac{2q}{q+2} \leq \mu < q$, <u>il existe une constante</u> A_μ <u>telle que, pour toute</u> $f \in L^\mu(K/M)$ <u>et pour tout</u> $H \in \mathfrak{a}^+$, <u>on ait</u> :

$$Tf(H) \leq A_\mu e^{\frac{2}{\mu}\rho(H)} \|f\|_\mu .$$

Démonstration. Soit ν tel que $\frac{1}{q} = \frac{1}{\mu} + \frac{1}{\nu} - 1$. Alors $1 < \nu \leq 2$. D'après l'inégalité de Young et le lemme 1, on a

$$\| \sqrt{P_H} * f \|_q \leq \|\sqrt{P_H}\|_\nu \|f\|_\mu \leq c_\nu\, e^{(1-\frac{2}{\nu})\,\rho(H)} \|f\|_\mu$$

donc

$$Tf(H) \leq c_\nu e^{(1+\frac{2}{q} + 1-\frac{2}{\nu})\,\rho(H)} \|f\|_\mu = c_\nu\, e^{\frac{2}{\mu}\rho(H)} \|f\|_\mu .$$

LEMME 3. <u>Considérons</u> T <u>définie en</u> (5) <u>comme une application sous-linéaire qui transforme les fonctions sur</u> K/M <u>(muni de la mesure</u> dk_M<u>)</u> <u>en fonctions sur</u> $\mathfrak{a}^+$ <u>(muni de la mesure</u> $d\sigma(H) = e^{-2\rho(H)}\,dH$). <u>Alors, pour tout</u> μ <u>tel que</u> $\frac{2q}{q+2} \leq \mu < q$, <u>l'application</u> T <u>est de type faible</u> (μ,μ) <u>au sens de Marcinkiewicz</u>

Démonstration. Pour tout $s > 0$, soit E_s l'ensemble des $H \in \mathfrak{a}^+$ tels que $Tf(H) > s$. Il faut montrer que $\sigma(E_s) = c\left(\frac{\|f\|_\mu}{s}\right)^\mu$. Or, si $H \in E_s$, d'après le lemme 2 on a

$$e^{\frac{2}{\mu}\rho(H)} \|f\|_\mu \geq A_\mu^{-1} s ,$$

donc
$$e^{-2\rho(H)} \leq A_\mu^\mu \left(\frac{\|f\|_\mu}{s}\right),$$

d'où le résultat.

LEMME 4. <u>Pour tout</u> $q > 2$, <u>il existe une constante</u> C'_q <u>telle que, pour toute</u> $f \in L^2(K/M)$, <u>on ait</u> :

$$\int_{\mathfrak{a}^+} \|\sqrt{P_H} * f\|_q^2 \, e^{\frac{4}{q}\rho(H)} \, dH \leq C'_q \|f\|_2^2 . \tag{6}$$

Démonstration. Soient des nombres réels μ_1 et μ_2 tels que

$$\frac{2q}{q+2} \leq \mu_1 < 2 < \mu_2 < q.$$

D'après le lemme 3, l'application T est de type faible (μ_1,μ_1) et (μ_2,μ_2). Donc, d'après le théorème de Marcinkiewicz, T est de type fort $(2,2)$. Autrement dit, il existe une constante C'_q telle que, pour toute $f \in L^2(K/M)$, on ait :

$$\int_{\mathfrak{a}^+} |Tf(H)|^2 \, e^{-2\rho(H)} \, dH \leq C'_q \|f\|_2^2 ,$$

ce qui, en remplaçant $Tf(H)$ par sa définition (5), n'est autre que (6).

Fin de la démonstration du Théorème 2. Prouvons la formule (4) à partir des Lemmes 2 et 4. On a d'après le Lemme 4,

$$\int_{\mathfrak{a}^+} \|\sqrt{P_H} * f\|_q^q \, e^{2\rho(H)} \, dH =$$

$$\int_{\mathfrak{a}^+} \|\sqrt{P_H} * f\|_q^2 \, e^{\frac{4}{q}\rho(H)} \, \|\sqrt{P_H} * f\|_q^{q-2} \, e^{(2-\frac{4}{q})\rho(H)} \, dH \leq$$

$$\leq C'_q \|f\|_2^2 \sup_{H \in \mathfrak{a}^+} \left(\|\sqrt{P_H} * f\|_q^{q-2} \, e^{(2-\frac{4}{q})\rho(H)}\right).$$

Or, d'après le lemme 2 où l'on fait $\mu = 2$,

$$\| \sqrt{P_H} * f\|_q^{q-2} = Tf(H)^{q-2}\, e^{-(1+\frac{2}{q})(q-2)\,\rho(H)}$$

$$\leq A_2^{q-2}\, e^{(q-2)\,\rho(H)}\, e^{-(1+\frac{2}{q})(q-2)\,\rho(H)}\, \|f\|_2^{q-2}$$

$$\leq A_2^{q-2}\, e^{-\frac{2}{q}(q-2)\,\rho(H)}\, \|f\|_2^{q-2} .$$

Donc

$$\|\sqrt{P_H} * f\|_2^{q-2}\, e^{(2-\frac{4}{q})\,\rho(H)} \leq A_2^{q-2}\, \|f\|_2^{q-2} ,$$

et, par suite,

$$\int_{\mathfrak{a}^+} \|\sqrt{P_H} * f\|_q^q\, e^{2\rho(H)}\, dH \leq C'_q\, A_2^{q-2}\, \|f\|_2^2\, \|f\|_2^{q-2} = C_q\, \|f\|_2^q .$$

CQFD.

(1) EYMARD, P. et LOHOUE, N. A paraître aux Ann. Ec. Norm. Sup.

(2) HELGASON, S. Bounded spherical functions on symetric spaces. Advances in Math. 3 (1969).

(3) HARISCH-CHANDRA Spherical functions on semi-simple Lie groups. I. Amer. Math. Soc. 80 (1958), 241-593.

(4) ZYGMUND, A. Trigonometric series. Cambridge Univ. Press. 2nd. ed. 1968.

Université de Paris-Sud

Mathématiques

Bâtiment 425

91405 - ORSAY

Diagonalisation du système de de Rham-Hodge au dessus d'un espace Riemannien homogène

Marie Paule Malliavin et Paul Malliavin

Le système de de Rham-Hodge au dessus d'une variété Riemannienne V ne se réduit pas à un système diagonal dans la cas non homogène ; en effet les formules de la moyenne associées s'obtiennent en résolvant des équations différentielles linéaires à coefficients variables le long des trajectoires de la diffusion relevée [3] . Dans le cas où l'espace V est homogène les équations différentielles associées sont alors à coefficient constants et se résolvent ainsi par des exponentielles $e^{\gamma t}$ sur une décomposition adaptée de l'espace.

On se propose d'explicité le plus possible cette diagonalisation, y compris le calcul des exposants γ . (cf Théorème 16).

Un théorème d'annulation de cohomologie au dessus de V se ramènera (cf théorème 11) ainsi entièrement:

a) à un calcul des exposants γ calcul qui dépendra uniquement d'opérations effectuées dans l'algèbre de Lie.

b) à la détermination de la borne supérieure du spectre d'opérateurs de Casimir opérant sur des fonctions équivariantes. Un théorème d'annulation sera alors équivalent à la résolution de b), problème qui relève des méthodes de l'analyse (cf [2] qui constitue une première tentative dans cette direction).

Soit G un groupe de Lie que l'on supposera unimodulaire, H un sous-groupe compact, V = G/H et $p : G \longrightarrow G/H$ la projection. On a par hypothèse

$$g = \mathfrak{h} \oplus \rho$$

où $[\mathfrak{h},\rho] \subseteq \rho$, $[\rho,\rho] \subseteq \mathfrak{h}$ et on suppose ρ muni d'une métrique euclidienne invariante sous l'action de Ad(H) ; alors V sera muni de la structure riemannienne homogène héritée de celle de ρ. Si W est un fibré G-homogène sur V, on note $\Lambda^s(V,W)$ l'espace des formes différentielles de degré s sur V à valeurs dans W. Si $\rho \in \Lambda^s(V,W)$ et $x \in V$, l'application $\mathbb{R}$-linéaire

$$\rho_x : \Lambda^s T_x(V) \longrightarrow W_x$$

détermine une application $\mathbb{C}$-linéaire

$$\rho_x \in \mathrm{Hom}_{\mathbb{C}}(\Lambda^s T_x(V)_{\mathbb{C}}, W_x)$$

$$\simeq \Lambda^s T_x(V)^*_{\mathbb{C}} \oplus W_x$$

Donc $\Lambda^s(V,W)$ est l'espace des sections du fibré vectoriel $\Lambda^s T(V)^*_{\mathbb{C}} \otimes_V W$.

On identifiera à ρ l'espace tangent en $\dot{e} = eH$ à V ; comme on a $Ad(h)\rho = \rho$ pour tout $h \in H$, ρ est l'espace d'une représentation de K par transformations orthogonales. On notera λ_o la représentation adjointe de H dans $\rho^*_{\mathbb{C}}$ et par λ^r_o la représentation de H dans $\Lambda^r \rho^*_{\mathbb{C}}$.

Si λ est une représentation unitaire irréductible de H dans l'espace vectoriel F et si F^λ est le fibré vectoriel G-homogène sur V associé, le fibré vectoriel G-homogène $\Lambda^r T(V)^*_{\mathbb{C}} \otimes_V F$ est défini par la représentation $\mu := \lambda^r_o \otimes \lambda$ de H dans

$$F_1 := \Lambda^r \rho^*_C \otimes F$$

Pour une représentation $\lambda, \mu, \ldots$ de H on notera $\underline{\lambda}$, $\underline{\mu}$, ... sa dérivée.

Nous noterons Δ_G l'opérateur différentiel défini par

$$\Delta_G := \sum_{k=1}^{q} \mathcal{L}^2_{e_k}$$

où $e_1, e_2, \ldots, e_q$ étant une base orthonormale de p, on note par $\mathcal{L}_{e_k}$ la dérivée de Lie par rapport au champ de vecteur défini par e_k.

Si donc $f : G \longrightarrow F_1$ est une fonction C^∞, on a :

$$(\Delta_G f)(g) = \sum_{k=1}^{q} \frac{d^2}{d\tau^2} \Big[f(g \exp (\tau e_k))\Big]_{\tau=o} .$$

<u>1. Proposition</u>. <u>L'opérateur</u> Δ_G <u>conserve</u> $\mathcal{C}^\mu(G,F_1)$; <u>i.e. si</u> $f \in \mathcal{C}^\mu(G,F_1)$ <u>alors</u> $\Delta_G f \in \mathcal{C}^\mu(G ; F_1)$.

<u>Preuve</u>. On a, puisque f est μ-équivariante,

$$(\Delta_G f)(gh) = \sum_{k=1}^{q} \frac{d^2}{d\tau^2} \mu(h^{-1}) f(g \exp (\tau \, Adh\, e_k))_{\tau=o} \quad \text{où } g \in G,\ h \in H.$$

Puisque $\mu(h^{-1})$ est une constante, il résulte du fait que Adh transforme une base orthonormale de ρ en une autre et du fait que Δ_G est indépendant du choix d'une telle base orthonormale que

$$(\Delta_G f)(gh) = \mu(h^{-1})(\Delta_G f)(g).$$

<u>2. Proposition</u>. <u>Soit</u> J <u>un endomorphisme du</u> $\mathbb{C}$<u>-espace vectoriel</u> F_1 . <u>On suppose que, pour chaque</u> $f \in \mathcal{C}^\mu(G ; F_1)$ <u>on a</u> $J f \in \mathcal{C}^\mu(G ; F_1)$.
<u>Alors</u> J <u>commute à la représentation</u> μ (ici la fonction $J.f : G \longrightarrow F_1$ est définie par

$$(Jf)(g) = J(f(g)))$$

<u>Preuve</u>. Si $f \in \mathcal{C}^\mu(G ; F_1)$ on a, par hypothèse,

$$J(\mu(h^{-1})f(g)) = \mu(h^{-1})\, J(f(g))$$

pour tout $g \in G$ et $h \in H$.

Pour en déduire que

$$J \circ \mu(h^{-1}) = \mu(h^{-1}) \circ J \quad \text{pour tout } h \in H \text{ il suffit de}$$

démontrer que tout vecteur $v \in F_1$ est l'image par une fonction μ-équivariante d'un élément de G.

Soit $v \in F_1$ et $g_o \in G$. Posons $x_o = p(g_o) \in V$. On construit $f \in \mathcal{C}^{\mu}(G ; F_1)$ telle que $f(g_o) = v$ de la manière suivante. Soit U une carte locale au voisinage de x_o et $\psi : U \longrightarrow G$ une section de $p : G \longrightarrow G/H$, $p \circ \psi = \mathrm{id}$, de sorte que

$$p^{-1}(U) \simeq H \times U.$$

Définissons $f_1 : p^{-1}(U) \longrightarrow F_1$ par $f_1(\gamma) = \mu(h^{-1})v$ si $\gamma = (h,x) \in H \times U$. Alors f_1 est $\mathcal{C}^{\infty}$ et on la prolonge à $f : G \longrightarrow F_1$ en posant $f = \ell \cdot f_1$ où ℓ est une fonction scalaire sur G prenant la valeur 1 en x_o et dont le support est contenu dans U. On a alors

$$f(g_o) = \ell(p(g_o)).f_1(g_o) = v .$$

On identifiera l'espace vectoriel $\Lambda^r(G/H ; F)$ à $\mathcal{C}^{\mu}(G ; F_1)$. Si $\omega \in \Lambda^r(G/H ; F)$, on notera f_ω la fonction équivariante correspondante

$$f_\omega(g) = \sum_H a_H(g) \zeta^H$$

où $H = (i_1, i_2, \ldots, i_r)$ est un multi indice ordonné. Alors

$$f_{d\omega} = \sum_{H,i} (\partial_i a_H) \zeta^i \wedge \zeta^H$$

où ∂_i est la dérivée de Lie $\mathcal{L}_{e_i}$ sur G.

On définit, si $\omega \in \Lambda^r(V ; F)$,

$$\|\omega_x\|^2 = \mathrm{Trace}\ \omega_x^* \omega_x$$

où l'adjoint ω_x^* est défini relativement à la structure hermitienne de $\rho_{\mathbb{C}} = \rho \otimes \mathbb{C}$ prolongeant la structure euclidienne de ρ. On pose

$$\|\omega\|_{L^2}^2 = \int_V \|\omega_x\|^2 \, dx$$

et on note $(\omega|\pi)$ le produit scalaire associé à cette norme c'est-à-dire :

$$(\omega|\pi)_{L^2} = \sum_H \int_G (a_H|b_H)_F \, dg$$

si $f_\omega = \sum_H a_H \zeta^H$ $\quad f_\pi = \sum_H b_H \zeta^H$

Comme G est unimodulaire on a le lemme :

<u>3. Lemme. Si</u> v_1, v_2 <u>sont deux fonctions scalaires définies sur</u> G <u>à support compact on a</u> :

$$\int_G (\partial_i v_1)(g)\, v_2(g)\, dg = -\int_G v_1(g)\, (\partial_i v_2)(g)\, dg$$

<u>4. Lemme. Soit</u> d^* <u>l'opérateur défini sur les formes à support compact par</u> :

$(d\omega|\pi)_{L^2} = (\omega|d^*\pi)_{L^2}$ <u>où</u>

$\omega \in \Lambda^r(V\,;\,F), \pi \in \Lambda^{r+1}(V\,;\,F)$. <u>Alors si</u> :

$$f_\pi = \sum_K b_K \zeta^K$$

<u>on a</u>

$$f_{d^*\pi} = \sum_H -\varepsilon(i,H)\,\partial_i\,(b_{H\cup\{i\}})\,\zeta^H$$

<u>où</u>

$\varepsilon(i,H) = 0$ si $i \in H$ et sinon $\varepsilon(i,H)\,\zeta^i \wedge \zeta^H = \zeta^{H\cup i}$

<u>Preuve.</u> On a :

$$f_{d\omega} = \sum_{i,H} (\partial_i\, a_H)\,\zeta^i \wedge \zeta^H = \sum_{i,H} \varepsilon(i,H)\,(\partial_i\, a_H)\,\zeta^{H\cup i} .$$

D'où

$$(f_{d\omega}|f_\pi) = \sum_{i,H} \varepsilon(H,i)\,(\partial_i\, a_H|b_{H\cup i}) .$$

Le lemme 3 s'étend aux fonctions à valeurs vectorielles dans l'espace hermitien F :

$$\int_G (\partial_i\, a_H\,|\,b_S)\,dg = -\int_G (a_H|\,\partial_i\, b_S)\,dg ,$$

ce qui démontre le lemme.

<u>5. Corollaire.</u> <u>Notons</u> $e_k \llcorner$ <u>le produit intérieur par le vecteur</u> e_k, <u>défini par</u>

$$e_k \llcorner \zeta^H = \begin{cases} \varepsilon(k,H-k)\,\zeta^{H-k} & \text{si } k \in H \\ 0 \text{ si } k \notin H. \end{cases}$$

<u>Alors on a</u> :

$$f_{d^*\pi} = -\sum_{i,K} (\partial_i\, b_K)\, e_i \llcorner \zeta^K .$$

Nous ne discuterons pas ici <u>pour quels sous complexes</u> on pourra avoir

$$d^2 = 0$$

On posera <u>par définition</u>

$$\square\omega = dd^* + d^* d$$

et on définit $J \in \mathrm{End}_{\mathbb{C}}(\Lambda^r p^*_{\mathbb{C}} \otimes F)$ de la manière suivante : Si $A \in \mathcal{L}_r(p_{\mathbb{C}}\,;\,F)$, espace des applications r-linéaires antisymétriques identifié à $F_1 = \Lambda^r p^*_{\mathbb{C}} \otimes F$, on pose :

(6) $$B = JA$$

où B est définie par :

(7) $$B(z_1,z_2,\ldots,z_r) = \sigma_{(z_1,\ldots,z_r)}\left\{\sum_{i=1}^{q} (\underline{\mu}([z_1,e_i]\,A)(e_i,z_2,\ldots,z_r)\right\}$$

où $z_1,z_2,\ldots,z_r \in p_{\mathbb{C}}$ et où $\sigma_{(z_1,\ldots,z_r)}$ désigne l'opérateur d'antisymétrisation relatif au r-uple $(z_1,z_2,\dots,z_r)$.

Par exemple si $r=1$, on obtient simplement :

$$B(z) = \sum_{i=1}^{q} (\mu([z,e_i])A)(e_i) =$$

$$= \sum_{i=1}^{q} \lambda[z,e_i]A(e_i) - \sum_{i=1}^{q} A([[z,e_i], e_i]) .$$

Si $r=q$, alors J est nulle car il en est ainsi de $B(z_1,e_2,\dots,e_q)$.

6. Théorème. (Formule de Weitzenböck). Soit $\pi \in \Lambda^r(V ; F^\lambda)$ alors on a

$$f_{\square\pi} = -\Delta_G f_\pi + Jf_\pi$$

où J a été définie en (6)

Preuve. Le calcul sera conduit en indices non ordonnées. On notera $\vec{i}$ le multi-indice $(i_1,i_2,\dots,i_r)$. On définit pour $z \in \rho_{\mathbb{C}}$, l'opérateur $z\llcorner$ sur $\Lambda\rho^*_{\mathbb{C}}$; $z\llcorner$ est $\mathbb{C}$-linéaire, c'est une anti-dérivation de degré -1 :

$$z\llcorner : \Lambda^r\rho^*_{\mathbb{C}} \longrightarrow \Lambda^{r-1}\rho^*_{\mathbb{C}}$$

vérifiant :

$$z\llcorner(u\wedge v) = (z\llcorner u)\wedge v + (-1)^{\deg u}\, u\wedge(z\llcorner v) \quad \text{pour } u,v \in \Lambda\rho^*_{\mathbb{C}}$$

Si $u \in \Lambda^1(\rho^*_{\mathbb{C}})$ $z\llcorner u = \langle z,u\rangle$, accouplement naturel de $\rho_{\mathbb{C}}$ et $\rho^*_{\mathbb{C}}$.

On remarquera que la définition de $z\llcorner$ coïncide avec celle de $e_j \llcorner \zeta^H$ (où H est ordonnée) de 5. Soit π une forme de degré r et soit

$$f_\pi = \frac{1}{r!} \sum_{\vec{i}} a_{\vec{i}}\, \zeta^{\vec{i}}$$

la fonction équivariante associée.

Alors

$$f_{d\pi} = \frac{1}{r!} \sum_{s,\vec{i}} \partial_s a_{\vec{i}}\, \zeta^s \wedge \zeta^{\vec{i}}$$

et

$$f_{d^*\pi} = -\frac{1}{r!} \sum_{p,\vec{i}} \partial_p a_{\vec{i}}\, e_p \llcorner \zeta^{\vec{i}}$$

$$f_{d^*d\pi} = -\frac{1}{r!} \sum_{t,s,\vec{i}} \partial_t \partial_s a_{\vec{i}}\, e_t \llcorner (\zeta^s \wedge \zeta^{\vec{i}})$$

$$f_{dd^*\pi} = -\frac{1}{r!} \sum_{k,p,\vec{j}} \partial_k \partial_p a_{\vec{j}}\, \zeta^k \wedge (e_p \llcorner \zeta^{\vec{j}})$$

On a par définition du produit intérieur $e_t\llcorner$:

$$e_t \llcorner (\zeta^s \wedge \zeta^{\vec{i}}) = \langle e_t, \zeta^s\rangle \zeta^{\vec{i}} - \zeta^r \wedge (e_t \llcorner \zeta^{\vec{i}})$$

$$= \begin{cases} \zeta^{\vec{i}} & \text{si } t=s \\ -\zeta^r \wedge (e_t \llcorner \zeta^{\vec{i}}) & \text{si } t\neq s \end{cases}$$

Dans la somme $f_{\square\pi} = f_{d^*d\pi} + f_{dd^*\pi}$, on regroupe les termes correspondants à $t=s$ et $k=p$ et cette dernière somme est égale à $-\Delta_G f_\pi$ car on a

$$e_t \llcorner (\zeta^t \wedge \zeta^{\vec{i}}) = \begin{cases} \zeta^{\vec{i}} & \text{si } t \notin \vec{i} \\ 0 & \text{sinon} \end{cases}$$

$$\zeta^p \wedge (z_p \llcorner \zeta^{\vec{j}}) = \begin{cases} \zeta^{\vec{j}} & \text{si } p \in \vec{j} \\ 0 & \text{sinon.} \end{cases}$$

D'où $f_{\square\pi} + \Delta_G f_\pi$ est égale à

$$-\frac{1}{r!} \sum_{\substack{\vec{i} \\ t\neq s}} \partial_t \partial_s a_{\vec{i}}\, e_t \llcorner (\zeta^s \wedge \zeta^{\vec{i}})$$

$$-\frac{1}{r!} \sum_{\substack{\vec{j} \\ k\neq p}} \partial_k \partial_p a_{\vec{j}}\, \zeta^k \wedge (e_p \llcorner \zeta^{\vec{j}})$$

$$= \frac{1}{r!} \sum_{\substack{\vec{i} \\ t\in\vec{i} \\ t\neq s}} (\partial_t \partial_s - \partial_s \partial_t)\, a_{\vec{i}}\, \zeta^s \wedge (e_t \llcorner \zeta^{\vec{i}})$$

ceci car $e_t \llcorner (\zeta^s \wedge \zeta^{\vec{i}}) = -\zeta^s \wedge (e_t \wedge \zeta^{\vec{i}})$ pour $t\neq s$.

Puisque f_π est équivalente,

$$(\partial_t \partial_s - \partial_s \partial_t)\, a_{\vec{i}}$$

est la composante $\vec{i}$-ième de $\underline{\mu}([e_s,e_t])\, f_\pi$ où $\underline{\mu}$ est la dérivée de la représentation μ ; c'est-à-dire

$$(\partial_t \partial_s - \partial_s \partial_t)\, a_{\vec{i}} = (\underline{\mu}([e_s,e_t])\, f_\pi)\, (e_{i_1}, e_{i_2}, \dots, e_{i_r}) \quad \text{si}$$

$\vec{i} = (i_1, i_2, \dots, i_r)$.

Explicitons la relation entre $\vec{k}$ et $\vec{i}$ où :

$$(-1)^{\alpha-1}\, \zeta^{\vec{k}} = \zeta^s \wedge (e_t \llcorner \zeta^{\vec{i}}).$$

Alors si $\vec{i} = (i_1, i_2, \dots, i_r)$ on a

$$\vec{k} = (s, i_1, \dots, i_{\alpha-1}, t, i_{\alpha+1}, \dots, i_r) \quad \text{où} \quad i_\alpha = t$$

D'où : $\vec{i} = (k_2, \dots, k_\alpha, t, k_{\alpha+1}, \dots, k_r)$ et

$$f_{\square\pi} + \Delta_G f_\pi = \frac{1}{r!} \sum_{\substack{\vec{i}=(k_2,\dots,k_\alpha,t,k_{\alpha+1},\dots,k_r) \\ t,k_1\neq t}} (-1)^{\alpha-1} (\underline{\mu}([e_{k_1}, e_t]\, f_\pi)_{\vec{i}}\, \zeta^{\vec{k}}$$

$$= \frac{1}{r!} \sum_{\substack{\vec{i}=(k_2,\dots,k_\alpha,t,k_{\alpha+1},\dots,k_r) \\ t,k_1\neq t}} \underline{\mu}([e_{k_1}, e_t]\, f_\pi)(e_t, e_{k_2}, \dots, e_{k_r})\, \zeta^{\vec{k}}.$$

D'où, en regroupant les termes qui correspondent au même multi indice $\vec{k}$ ordonné, on a si, $Jf_\pi = h$:

9. Lemme. La matrice J définie en (6) commute avec la représentation μ et, pour la structure hermitienne de $\Lambda^r \rho^*_{\mathbb{C}} \otimes F$, elle est hermitienne.
(On munit $\rho^*_{\mathbb{C}}$ de la structure hermitienne induisant sur ρ^* la structure euclidienne donnée).

Preuve. La première partie du lemme résulte de 2. Puisqu'il s'écrit sous la forme $dd^* + d^*d$, $\Box$ est hermitien. Δ_G est hermitien en vertu de 3. Par suite, à cause de la surjectivité $\Lambda^r(V ; F^\lambda) \longrightarrow \mathcal{E}^\mu(G ; F_1)$, $J = \Box + \Delta_G$ est hermitien.

10. Spectres.

Spectre de la matrice J.

Si θ est une représentation unitaire de H on note $\mathcal{E}(\theta)$ les classes d'isotypie des sous représentations irréductibles contenues dans θ .

On décompose l'espace $\Lambda^r \rho^*_{\mathbb{C}} \otimes F$ par les projecteurs

$$P_\sigma = \int_H \mu(h)\, \overline{\chi(h)}\, dh \quad , \quad \sigma \in \mathcal{E}(\mu) \; .$$

Comme J commute avec μ, J conserve l'image de P_σ ; on note pour J_σ la restriction de J à cette image . J_σ est un endomorphisme hermitien diagonalisable, on considère son spectre $sp(J_\sigma)$ et on posera

$\gamma(\sigma, r, \lambda) = \inf(sp(J_\sigma))$.

Spectee du laplacien horizontal.

Si σ est une représentation irréductible de H, nous poserons
$L^2_\sigma(G) = \mathcal{E}^\sigma \cap L^2(G)$ $\quad sp(\sigma) = \{ \lambda \in \mathbb{R}^+ \;;$ il existe $f \in L^2_\sigma(G) : \Delta_G f + \lambda f = 0. \}$
Si $\inf(sp(\sigma)) \in sp(\sigma)$ nous posons $C(\sigma) = \inf(sp(\sigma))$. Sinon nous poserons $C(\sigma) = +\infty$.

11. Théorème de scalarisation.

Donnons nous une représentation unitaire λ de H alors le système

$$\begin{cases} \Box\omega = 0 \\ \omega \in \Lambda^r(G/H, F^\lambda) \;, \omega \in L^2 \end{cases}$$

entraîne $\omega = 0$
si et seulement si on a $\quad C(\sigma) + \gamma(\sigma, r, \lambda) > 0$ pour tout $\sigma \in \mathcal{E}(\chi^r_o \otimes \lambda)$

Preuve. Introduisons la fonction équivariante associée f_ω et posons $P_\sigma f_\omega = f^\sigma_\omega$. Le système s'écrit

$$\begin{cases} -\Delta_G f^\sigma_\omega + J_\sigma f^\sigma_\omega = 0 \qquad \sigma \in \mathcal{E}(\mu) \\ f^\sigma_\omega \in L^2(G). \end{cases}$$

J_σ commute encore avec μ ; ainsi chaque sous espace propre de J_σ est invariant sous l'action de μ. Notons $f^{\sigma,\xi}_\omega$ la composante de f^σ_ω sur le sous espace propre de J_σ associé à $\xi \in sp(J_\sigma)$ le système s'écrit

$$\begin{cases} -\Delta_G f_\omega^{\sigma,\xi} + \xi f_\omega^{\sigma,\xi} = 0 & \sigma \in \mathcal{E}(\mu), \xi \in sp(J_\sigma) \\ f_\omega^{\sigma,\xi} \in L^2(G) & . \end{cases}$$

Ainsi l'existence de $\omega \neq 0$ est équivalente à

$$(-sp(J_\sigma)) \cap sp(\sigma) \neq \emptyset \quad \text{pour un } \sigma \in \mathcal{E}(\mu) .$$

D'autre part l'opérateur $\square$ est <u>positif</u>, par suite pour tout $\xi \in sp(J_\sigma)$

$$(-\Delta_G + \xi) \text{ est positif sur } L^2_\sigma$$

d'où

$$sp(\sigma) \subset [-\gamma(\sigma, r, \lambda), +\infty[$$

L'intersection de $(-sp(J_\sigma))$ et de $sp(\sigma)$ ne peut être ainsi que le point $\gamma(\sigma, r, \lambda)$ et le théorème est établi.

A partir de maintenant, on suppose que G est un groupe de Lie semi-simple connexe à centre fini et que H est un sous-groupe compact maximal. On munit ρ de la structure euclidienne induite par la forme de Killing B. On choisit une base $(e_{q+1},\dots,e_n)$ de $\mathfrak{h}$ de telle sorte que

$$B(e_\alpha, e_\beta) = -\delta_{\alpha\beta} \qquad q+1 \leq \alpha, \beta \leq n.$$

On a alors :

<u>12. Lemme</u>. <u>Si</u> $u \in \mathcal{E}^\mu(G\,;\,F_1)$ <u>on a</u> :

$$Ju = -\frac{1}{r} \sum_{\alpha=q+1}^{n} (\underline{\mu}(e_\alpha)u) \circ \underline{\lambda_o^r} (e_\alpha)$$

<u>Preuve</u>. On a, pour $j_1, j_2, \dots, j_r$ compris entre 1 et q,

$$(Ju)(e_{j_1} \wedge e_{j_2} \wedge \dots \wedge e_{j_r})$$

$$= \sigma_{(j_1,\dots,j_r)} \sum_{i=1}^{q} (\underline{\mu}([e_{j_1}, e_i])u)(e_i \otimes e_{j_2} \otimes \dots \otimes e_{j_r}) .$$

On a :

$$[e_{j_1}, e_i] = \sum_{\alpha=q+1}^{n} c^\alpha_{j_1,i} \; e_\alpha \quad \text{où les } c^\alpha_{j_1,i}$$

sont des constantes de structure de $\mathfrak{g}$ (cf (4)).

D'où

$$(Ju)(e_{i_1} \wedge \dots \wedge e_{j_r})$$

$$= \sigma_{(j_1,\dots,j_r)} \left\{ \sum_{\alpha=q+1}^{n} (\underline{\mu}(e_\alpha)u)(\sum_{i=1}^{q} c^\alpha_{j_1,i} \; e_i \otimes e_{j_2} \otimes \dots \otimes e_{j_r}) \right. .$$

Mais

$$\sum_{i=1}^{q} c^\alpha_{j_1,i} \; e_i = \sum_{i=1}^{q} c^i_{j_1,\alpha} \; e_i = [e_{j_1}, e_\alpha] \text{ (cf [4])}$$

Donc : $(Ju)(e_{i_1}\wedge\ldots\wedge e_{j_r})$

$$=\sigma_{(j_1,\ldots,j_r)}\left\{\sum_{\alpha=q+1}^{n}(\underline{\mu}(e_\alpha)u)([e_{j_1},e_\alpha]\otimes e_{j_2}\otimes\ldots\otimes e_{j_s})\right\}$$

$$=-\sum_{\alpha=q+1}^{n}(\underline{\mu}(e_\alpha)u)\quad(\sigma_{(j_1,\ldots,j_r)}\ \underline{\lambda}_o(e_\alpha)(e_{j_1})\otimes e_{j_2}\otimes\ldots\otimes e_{j_r}\} .$$

Par définition.

$$\sigma_{(j_1,\ldots,j_r)}\left\{\underline{\lambda}_o(e_\alpha)(e_{j_1})\otimes e_{j_2}\otimes\ldots\otimes e_{j_r}\right\}$$

$$=\frac{1}{r!}\sum_{\sigma\in S_r}\varepsilon(\sigma)\ \underline{\lambda}_o(e_\alpha)(e_{\sigma(j_1)})\otimes e_{\sigma(j_2)}\otimes\ldots\otimes e_{\sigma(j_r)}$$

où S_r est le groupe symétrique de degré r et $\varepsilon(\sigma)$ la signature de la permutation σ .

Comme $v=\underline{\mu}(e_\alpha)u$ est r-antilinéaire, on a :

$$v(\sigma_{(j_1,\ldots,j_r)}\left\{\underline{\lambda}_o(e_\alpha)(e_{j_1})\otimes e_{j_2}\otimes\ldots\otimes e_{j_r}\right\})$$

$$=\frac{1}{r!}\sum_{\sigma\in S_r}\varepsilon(\sigma)\ v\left\{\underline{\lambda}_o(e_\alpha)(e_{\sigma(j_1)})\wedge e_{\sigma(j_2)}\wedge\ldots\wedge e_{\sigma(j_r)}\right\} .$$

Notons S_{r-1} le sous-groupe de S_r formé des $\sigma\in S_r$ tels que $\sigma(j_1)=j_1$. Considérons la partition $S_r=\bigcup_{i=1}^{r}S_{r-1}\tau_i$ où τ_i est la transposition $j_1\longleftrightarrow j_i$ $i=1,2,\ldots,r$ et où par conséquent τ_1 est l'identité.

On a alors :

$$(*)\quad v(\sigma_{(j_1,\ldots,f_r)}\left\{\underline{\lambda}_o(e_\alpha)(e_{j_1})\otimes e_{j_2}\otimes\ldots\otimes e_{j_r}\right\})$$

$$=\frac{1}{r}\sum_{i=1}^{r}\left\{\frac{1}{(r-1)!}\sum_{\sigma\in S_{r-1}\tau i}\varepsilon(\sigma)\ v(\underline{\lambda}_o(e_\alpha)(e_{\sigma(j_1)})\wedge e_{\sigma(j_2)}\wedge\ldots\wedge e_{\sigma(j_r)}\right).$$

Le premier terme entre accolades vaut :

$$\frac{1}{(r-1)!}\sum_{\sigma\in S_{r-1}}\varepsilon(\sigma)\ v(\underline{\lambda}_o(e_\alpha)(e_{\sigma(j_1)})\wedge e_{\sigma(j_2)}\wedge\ldots\wedge e_{\sigma(j_r)})$$

$$=v(\underline{\lambda}_o(e_\alpha)(e_{j_1})\wedge\left[\frac{1}{r-1}\sum_{\sigma\in S_{r-1}}\varepsilon(\sigma)\ e_{\sigma(j_2)}\wedge\ldots\wedge e_{\sigma(j_r)}\right])$$

$=v(\underline{\lambda}_o(e_\alpha)(e_{j_1})\wedge e_{j_2}\wedge\ldots\wedge e_{j_r})$ car l'opérateur d'antisymétrisation est un projecteur.

Pour $i\geqslant 2$, le terme entre accolade correspondant à l'indice i dans l'expression (*) s'écrit :

$$(**) = \frac{1}{(r-1)!} \sum_{\sigma\tau_i \in S_{r-1}} -\varepsilon(\sigma\tau_i) v(\underline{\lambda}_o(e_\alpha)(e_{\sigma\tau_{i(j_i)}}) \wedge e_{\sigma\tau_{i(\tau_i j_2)}} \wedge \dots \wedge e_{\sigma\tau_{i(\tau_i j_r)}})$$

$$= \frac{1}{(r-1)!} \sum_{\sigma \in S_{r-1}} -\varepsilon(\sigma) v \left[\underline{\lambda}_o(e_\alpha)(e_{\sigma(j_2)}) \wedge e_{\sigma(\tau_i j_2)} \wedge \dots \wedge e_{\sigma(\tau_i j_i)} \wedge \dots \wedge e_{\sigma(\tau_i j_r)}\right) .$$

Mais $\tau_i j_i = j_1$ et $\sigma(j_1) = j_1$ pour $\sigma \in S_{r-1}$. Or

$$\underline{\lambda}_o(e_\alpha)(e_{\sigma(j_i)}) \wedge e_{\sigma(\tau_i j_2)} \wedge \dots \wedge e_{\sigma(\tau_i j_i)} \wedge \dots \wedge e_{\sigma(\tau_i j_r)}$$

$$= - e_{\sigma(\tau_i j_i)} \wedge e_{\sigma(\tau_i j_2)} \wedge \dots \wedge \underline{\lambda}_o(e_\alpha)(e_{\sigma(j_i)}) \wedge \dots \wedge e_{\sigma(\tau_i j_r)} .$$

Donc (**) =

$$\frac{1}{(r-1)!} \sum_{\sigma \in S_{r-1}} \varepsilon(\sigma)\sigma \left[e_{j_1} \wedge e_{\sigma(j_2)} \wedge \dots \wedge \underline{\lambda}_o(e_\alpha)(e_{\sigma(j_i)}) \wedge \dots \wedge e_{\sigma(j_r)}\right)$$

$$= v(e_{j_1} \wedge e_{j_2} \wedge \dots \wedge \underline{\lambda}_o(e_\alpha)(e_{j_i}) \wedge \dots \wedge e_{j_r}).$$

Par suite.

$$Ju = -\frac{1}{r} \sum_{\alpha=q+1}^{n} (\underline{\mu}(e_\alpha)u) \circ \underline{\lambda}_o^r (e_\alpha).$$

<u>13 Proposition</u>. On a :

$$Ju = -\frac{1}{2r} (\underline{\mu}(\Delta_H)u) + \frac{1}{2r} \underline{\lambda}(\Delta_H) \circ u - \frac{1}{2r} u \circ \underline{\lambda}_o^r (\Delta_H)$$

<u>où</u> Δ_H <u>est le casimir de</u> $\mathfrak{h}$.

Cette proposition résultera des deux lemmes suivants :

<u>14 Lemme</u>. <u>Si</u> ν <u>et</u> τ <u>sont deux représentations de</u> H <u>et si</u> $\theta = \nu \otimes \tau$ <u>on a</u>

$\underline{\theta}(\Delta_H) - \underline{\nu}(\Delta_H) \otimes 1_\tau - 1_\nu \otimes \underline{\tau}(\Delta_H) = 2 \sum_{\alpha=q+1}^{n} \underline{\nu}(e_\alpha) \otimes \underline{\tau}(e_\alpha)$ <u>où</u>

1_τ , 1_ν <u>représentent les applications identités de l'espace de la représentation</u> τ <u>et</u> ν <u>respectivement et où</u> $(e_\alpha)_{q+1 \leq \alpha \leq n}$ <u>est une base orthonormale de</u> $\mathfrak{h}$.

<u>Preuve.</u> Dérivons suivant $\mathcal{L}_{e_\alpha}$ deux fois l'expression

$$\nu(h) \otimes \tau(h) \qquad h \in H$$

$$\mathcal{L}^2_{e_\alpha} (\nu(h) \otimes \tau(h)) = \nu(h) (\underline{\nu}(e_\alpha))^2 \otimes \tau(h)$$

$$+ 2\nu(h)\underline{\nu}(e_\alpha) \otimes \tau(h)\underline{\tau}(e_\alpha) + \nu(h) \otimes \tau(h)(\underline{\tau}(e_\alpha))^2 .$$

D'où pour h=e élément unité de H on a :

$$(\mathcal{L}^2_{e_\alpha}(\nu \boxtimes \tau))_{h=e} = (\underline{\nu}(e_\alpha))^2 \boxtimes 1_\tau + 1_\nu \boxtimes (\underline{\tau}(e_\alpha))^2 + 2\underline{\nu}(e_\alpha) \boxtimes \underline{\tau}(e_\alpha) .$$

D'où le résultat en sommant sur $\alpha = q+1,\dots,n$.

Il résulte du lemme précédent que :

$$Ju = -\frac{1}{r} \sum_{\alpha=q+1}^{n} \underline{\lambda}(e_\alpha) \circ u \circ \underline{\lambda}_o^r(e_\alpha) - \frac{1}{r} \sum_{\alpha=q+1}^{n} u \circ (\underline{\lambda}_o^r(e_\alpha))^2$$

et on applique :

<u>15. Lemme. On a</u> :

$$2 \sum_{\alpha=q+1}^{n} \underline{\lambda}(e_\alpha) \circ u \circ \underline{\lambda}_o^r(e_\alpha) = \underline{\mu}(\Delta_H)\, u - \underline{\lambda}(\Delta_H) \circ u - u \circ \underline{\lambda}_o^r(\Delta_H) .$$

<u>Preuve</u>. Posons

$$\varphi(\lambda(h), \lambda_o^s(h)) = \lambda(h) \circ u \circ \lambda_o^r(h).$$

Alors φ est une application bilinéaire :

$$\varphi : \mathrm{End}\,(F) \times \mathrm{End}\,(\Lambda^r \rho_{\mathbb{C}}) \longrightarrow \mathrm{End}\,(F_1)$$

et l'on a :

$$\varphi(\lambda(h), \lambda_o^r(h))' = \varphi(\lambda'(h), \lambda_o^r(h)) + \varphi(\lambda(h), (\lambda_o^r)'(h)).$$

D'où en dérivant deux fois par rapport à $\mathcal{L}_{e_\alpha}$

$$\mathcal{L}^2_{e_\alpha} \varphi(\lambda(h), \lambda_o^r(h)) = \varphi(\lambda(h)\underline{\lambda}(e_\alpha)^2, \lambda_o^r(h))$$
$$+ 2\varphi(\lambda(h)\underline{\lambda}(e_\alpha), \lambda_o^r(h)\underline{\lambda}_o^r(e_\alpha)) + \varphi(\lambda(h), \lambda_o^r(h)\underline{\lambda}_o^r(e_\alpha)).$$

D'où pour h=e élément unité de H

$$(\mathcal{L}^2_{e_\alpha} \varphi(\lambda, \lambda_o^r))_{h=e} = \varphi(\lambda(e_\alpha)^2, 1) + 2\varphi(\underline{\lambda}(e_\alpha), \underline{\lambda}_o^r(e_\alpha)) + \varphi(1, \underline{\lambda}_o^r(e_\alpha))$$

et en sommant sur e_α $\alpha = q+1,\dots,n$.

$$\underline{\mu}(\Delta_H)u = \underline{\lambda}(\Delta_H)u + u \circ \underline{\lambda}_o^r(\Delta_H) + 2\, r \sum_{\alpha=q+1}^{n} \underline{\lambda}(e_\alpha) \circ u \circ \underline{\lambda}_o^r(e_\alpha) .$$

<u>16 Théorème</u>.

$$2\, r\gamma(\sigma, r, \lambda) = \underline{\lambda}(\Delta_H) - \sup_{\eta} \{\underline{\eta}(\Delta_H)\} + \underline{\sigma}(\Delta_H)$$

où $\sigma \in \mathcal{E}(\eta \boxtimes \lambda)$ et où $\eta \in \mathcal{E}(\lambda_o^r)$

<u>Remarque</u> : Le calcul peut ensuite être explicité à l'aide de la formule de Kostant déterminant $\mathcal{E}$.

<u>Preuve</u>.

Notons P^{η}_{σ} (resp P_{σ}) le projecteur sur la composante isotypique de type σ apparaissant dans $\lambda \otimes \eta$, (resp $\lambda \otimes \lambda^{r}_{o}$). Alors

$$\text{Image } (P_{\sigma}) = \bigoplus_{\eta} \text{ Image } (P^{\eta}_{\sigma}) \text{ .}$$

D'après 13, si $u \in P^{\eta}_{\sigma}$

$$2r\, Ju = (-\underline{\sigma}(\Delta_H) + \underline{\lambda}(\Delta_H) - \underline{\eta}(\Delta_H))\, u \text{ .}$$

Cette décomposition en somme directe de $\mathrm{Im}(P_{\sigma})$ diagonalise J_{σ} dont le spectre est mis en évidence le théorème est ainsi établi.

Institut Henri Poincaré
11, rue Pierre et Marie Curie
PARIS 5ème

<u>BIBLIOGRAPHIE</u>

1 P. Griffith et W.Schmid. Locally homogeneous complex manifold. Acta Mathematica 123 1969, p.252-302.

2 Marie Paule et Paul Malliavin . Spectre de l'opérateur de Casimir d'un groupe de Lie semi-simple et théorèmes d'annulation de cohomologie au dessus d'un espace Riemannien symétrique C.R. Acad. Sciences t. 279 (5 Aout 1974) p. 185-188.

3 Paul Malliavin (Novembre 1974). Formules de la moyenne et théorèmes d'annulation. Journal of Functional Analysis Novembre 1974.

4 Y. Matsushima. On the first betti number of compact quotient spaces of higher dimensional symetric spaces. Annals of mathematics. Vol 75 n°2 Mars 1962, p. 312-330.

ACTION DE CERTAINS GROUPES DANS DES ESPACES DE FONCTIONS C^∞

M. Rais

1. Introduction.

Les énoncés qu'on trouvera ci-dessous peuvent être considérés comme des réponses à une question (trop abstraite) du type suivant : Soit G un sous-groupe de $GL(R^n)$. Ce groupe opère naturellement dans divers espaces de fonctions définies sur R^n, par exemple l'espace S des fonctions polynômes, ou bien l'espace $\mathcal{E}$ des fonctions C^∞. Que peut-on dire alors de la représentation de G ainsi obtenue ? Je me suis placé successivement dans deux situations particulières :

(i) G est un groupe fini engendré par des réflexions

(ii) G est le groupe adjoint d'une algèbre de Lie compacte

où la représentation de G dans S est bien connue (respectivement Chevalley [2] et Kostant [5]). Il apparaitra ci-dessous que dans ces deux cas, la représentation de G dans $\mathcal{E}$ ne diffère pas essentiellement de celle de G dans S.

2. Groupe finis engendrés par des réflexions.

2.1. Soit V un espace vectoriel de dimension finie sur k (le corps k est soit celui R des nombres réels, soit celui C des nombres complexes). On désigne par S l'algèbre des fonctions polynômes sur V (à valeurs dans k), et par J une sous-algèbre de S ayant la propriété suivante : S est entière sur J. On sait qu'alors J est une algèbre de type fini. On se donne donc une famille ($u_1, \ldots, u_m$) d'éléments de J qui engendrent J comme k-algèbre, ce qui permet de définir une application $u : V \longrightarrow k^m$, dont les coordonnées sont $u_1, \ldots, u_m$. Alors l'application u est propre.

2.2. Soit maintenant W un sous-groupe fini de $GL(V)$ engendré par des réflexions, et soit J la sous-algèbre de S constituée par les éléments W-invariants. Soit $(u_1, \ldots, u_m)$ un système minimal de générateurs homogènes de l'idéal J_+ de J constitué par les polynômes sans terme constant. Alors on sait que m est la dimension de V, que $u_1, \ldots, u_m$ sont algébriquement indépendants, et que J est l'algèbre (de polynômes) engendrée par $u_1, \ldots, u_m$ ([1], chap.5, n° 5.3, théorème 3).

2.3. Ceci étant, on va utiliser ci-dessous la même lettre E pour désigner deux objets dont la définition est différente suivant que le corps de base est R ou C. Dans le premier cas, E est l'algèbre $\mathcal{E}(V)$ des fonctions C^∞ sur V et dans le deuxième E est l'algèbre $Hol(V)$ des fonctions holomorphes entières sur V. Si plus particulièrement $V = k^m$, on écrira E_m au lieu de E. On introduit l'application propre $u : V \longrightarrow k^m$ et l'application $u^* : E_m \longrightarrow E$ qui remonte sur V les fonctions définies sur k^m (on a $u^*(f) = f \circ u$). Cette dernière application est une application linéaire continue d'un Fréchet dans un autre, et en fait c'est un homomorphisme. Lorsque $k = C$, la démonstration est immédiate (si une suite de fonctions holomorphes sur C^m donne par u^* une suite qui converge uniformément vers zéro sur tout compact, elle converge uniformément vers zéro sur tout compact de l'image de u, du fait que u est propre, mais u est surjective). Lorsque $k = R$, celà résulte du théorème de Glaeser ([3]) et la démonstration est beaucoup plus difficile. On en déduit que l'image de u^* est l'algèbre E_o des fonctions W-invariantes.

2.4. Soit L un sous-espace vectoriel gradué W-invariant de S, supplémentaire dans S de l'idéal $J_+ S$ de S engendré par J_+. La représentation de W dans L est équivalente à la représentation régulière de W et l'application bilinéaire $(f,g) \longrightarrow fg$ de $J \times L$ dans S induit un isomorphisme de W-modules de $J \otimes L$ sur S ([1],chap.5,n° 5.2 théorème 2). S est donc un module libre sur J. De façon plus précise, si $(l_1, \ldots, l_p)$ est une base de L sur k constituée de polynômes homogènes, c'est aussi une base de S sur J. On a alors :

2.5. Proposition. E est un module libre sur E_o. Plus précisément, toute base de S sur J est une base de E sur E_o.

Soit maintenant F un sous-espace de E. On choisit une base $(l_1, \ldots, l_p)$ de E sur E_o. Si f est dans F, on sait qu'il existe des f_i dans E_o telles que $f = f_1 l_1 + \ldots + f_p l_p$. La réponse à la question suivante : Les f_i sont-elles dans F est affirmative lorsque F a les propriétés suivantes : F est stable par W, par les multiplications par les polynômes, et chaque fois qu'une fonction de F est divisible dans E par un polynôme, elle est divisible dans F par ce polynôme. Lorsque k est R, il en est ainsi de l'espace $\mathcal{D}(V)$ des fonctions à support compact, et aussi de l'espace $\mathcal{S}(V)$ des fonctions à décroissance rapide. Supposons maintenant $k = C$. Un premier exemple d'espace F est l'espace Exp(V) des fonctions holomorphes de type exponentiel. Un deuxième exemple est présenté dans le paragraphe suivant.

2.6. On part d'un espace vectoriel réel V' muni d'un groupe fini d'automorphismes W engendré par des réflexions, et on considère le dual complexe V de V'. Le groupe W opère naturellement dans V et il est engendré par des réflexions. L'espace $\mathcal{F}\mathcal{D}(V')$ des transformées de Fourier des fonctions C^∞ à support compact sur V' est un sous-espace F de Hol(V), de même que l'espace $\mathcal{F}\mathcal{E}'(V')$ des transformées de Fourier des distributions à support compact sur V'. L'énoncé suivant est alors évident.

2.7. Proposition. Soient V', V, et W comme indiqué. Soient $m_1, \ldots, m_p$ des distributions à support 0 dans V' et $d_1, \ldots, d_p$ les opérateurs différentiels à coefficients constants sur V' définis par $m_1, \ldots, m_p$. On suppose que la famille $(l_1, \ldots, l_p)$ des transformées de Fourier de $m_1, \ldots, m_p$ est une base de S sur J. On a alors : Soit f dans $\mathcal{D}(V')$. Il existe une unique famille $f_1, \ldots, f_p$ de fonctions dans $\mathcal{D}(V')$ W-invariantes telle que $f = d_1 f_1 + \ldots + d_p f_p$.

La proposition signifie en particulier que le plus petit sous-espace de $\mathcal{D}(V')$ qui contient les fonctions W-invariantes et leurs dérivées est $\mathcal{D}(V')$ lui-même. Sous cette forme, celà peut servir (comme me l'a signalé M. Duflo) à déterminer l'image de certaines transformations de Fourier attachées aux groupes semi-simples complexes.

2.8. Une dernière remarque concernant le cas des groupes finis : Soit W' un sous-groupe de W , et soit E' la sous-algèbre de E constituée par les fonctions W'-invariantes. Alors E' est un module libre sur E_o , de rang égal à l'indice de W' dans W .

3. Groupes adjoints d'une algèbre de Lie.

3.1. Dans ce qui suit, on désigne par $\mathfrak{g}$ une algèbre de Lie sur k , qui est compacte lorsque k = R , réductive lorsque k = C (on passe de la première situation à la deuxième par complexification) , et par G le groupe adjoint de $\mathfrak{g}$. L'algèbre S des fonctions polynômes sur $\mathfrak{g}$ est un G-module dont la structure est bien connue ([5]). Tout d'abord, on sait que l'algèbre J des fonctions polynômes G-invariants est une algèbre de polynômes de dimension m , où m est le rang de $\mathfrak{g}$ .Soit (u_1 ,..., u_m) un système de générateurs homogènes algébriquement indépendants de J et soit u l'application polynomiale de $\mathfrak{g}$ dans k^m associée. Lorsque k = R , cette application est propre ([6] , lemme 2 , chap.2). Elle ne l'est pas lorsque k = C , mais cependant admet une section holomorphe ([5] , theorem 7). Dans les deux cas, il vient que $u^* : E_m \longrightarrow E$ est un homomorphisme et que l'image de u* est la sous-algèbre E_o des fonctions G-invariantes. On peut signaler de plus que si on fixe une sous-algèbre de Cartan $\mathfrak{h}$ de $\mathfrak{g}$, l'opération de restriction à $\mathfrak{h}$ des fonctions définies dans $\mathfrak{g}$ induit un isomorphisme d'algèbres topologiques de E_o sur l'algèbre des fonctions invariantes par le groupe de Weyl de ($\mathfrak{g}$, $\mathfrak{h}$) .

3.2. Soit L le sous-espace vectoriel gradué de S constitué par les polynômes G-harmoniques ([4] , [5]). On a alors $S = L \oplus (J_+ S)$ et S est isomorphe comme G-module à $L \otimes J$. Soit δ une représentation irréductible (de dimension finie) de G . La multiplicité de δ dans L est finie. Soit L_δ l'espace vectoriel (de dimension finie) des polynômes harmoniques de type δ , et soit E_δ l'espace des fonctions de E qui sont de type δ .

3.3. Proposition. Chaque E_δ est un E_o-module libre. Plus précisément, toute base de L_δ (sur k) est une base de E_δ sur E_o .

L'espace $\sum E_\delta$ des fonctions G-finies est donc un module libre sur E_o , et toute base de L en est une base.

3.4. Supposons désormais que le corps de base est le corps des nombres réels, et désignons par $\mathfrak{g}_C$ la complexifiée de $\mathfrak{g}$. Les remarques de 2.6 se transposent de la manière suivante :

3.5. Proposition. Soit δ une représentation irréductible de G . Soient m_1 ,..., m_r des distributions à support $\{0\}$ dans $\mathfrak{g}$ et d_1 ,..., d_r les opérateurs différentiels associés On suppose que la famille (l_1 ,..., l_r) des transformées de Fourier de m_1 ,..., m_r est une base de $L_\delta(\mathfrak{g}_C)$. On a alors : Soit f une fonction dans $\mathcal{D}(\mathfrak{g})$, de type δ . Il existe une unique famille (f_1 ,.., f_r) de fonctions G-invariantes appartenant à

$\mathcal{D}(\mathfrak{g})$ telle que $f = d_1 f_1 + \dots + d_r f_r$.

En particulier, toute fonction G-finie est une somme finie de dérivées de fonctions invariantes.

BIBLIOGRAPHIE

[1] N. Bourbaki , Groupes et algèbres de Lie , chapitres 4,5 et 6,Hermann,Paris 1968.

[2] C. Chevalley, Invariants of a finite group generated by reflexions,Amer.J.Math.77 (778-782),1955.

[3] G. Glaeser , Fonctions composées différentiables,Ann. of Math.77,(193-209),1963.

[4] S. Helgason, Invariants and fundamental functions,Acta Math.109,(241-258),1963.

[5] B. Kostant , Lie group representation on polynomial rings,Amer.J.Math.85,(327-404) 1963.

[6] M. Rais, Distributions homogènes sur des espaces de matrices,Bull.Soc.Math. France,Mémoire n° 30 , 1972 .

Université de Poitiers
Département de Mathématiques
40, avenue du Recteur Pineau

86022 - POITIERS

MODELE DE WHITTAKER ET CARACTERES DE REPRESENTATIONS

F. R O D I E R

TABLE DES MATIERES

I - Introduction

Les fonctions de Whittaker ont été utilisées par H. Jacquet et R.P. Langlands [13], puis par J.A. Shalika [17] pour étudier l'espace des formes paraboliques sur GL (2), puis sur GL(n) . Ici nous étudions certaines propriétés des fonctions de Whittaker, en particulier la relation entre le caractère d'une représentation admissible et l'existence d'un modèle de cette représentation dans l'espace des fonctions de Whittaker. Pour cela, on construit une suite de représentations dont la limite est précisément l'espace des fonctions de Whittaker. Le point essentiel consiste à montrer que la limite inductive ainsi obtenue est stricte. On a utilisé pour cela la théorie de Kirillov sur les groupes compacts p-adiques, qui a été développée par R. Howe [12]. Cela nous a amené à faire quelques restrictions sur le corps de base.

Je dois ici exprimer ma dette à l'égard de R. Howe. Je me suis largement inspiré de sa conférence à Albany (S.U.N.Y., 1973) sur les fonctions de Whittaker sur GL(n), ainsi que des notes qu'il m'a permis de consulter.

II - Notations et préliminaires.

II.1. Soit K un corps localement compact, non discret et non archimédien. On notera $\mathcal{O}$ l'anneau des entiers de K, $\mathfrak{p}$ l'idéal maximal de $\mathcal{O}$, ϖ une uniformisante, p la caractéristique résiduelle de K, $|\ |_K$ une valeur absolue non triviale sur K, v_K une valuation de K, normalisée par la condition $v_K(\varpi) = 1$, $\overline{K}$ une clôture algébrique de K, et q le nombre d'éléments de $\mathcal{O}/\mathfrak{p}$.

Soit $\underline{\underline{G}}$ un groupe algébrique réductif connexe déployé sur K, $\underline{\underline{H}}$ un tore maximal de $\underline{\underline{G}}$ déployé sur K, $\underline{\underline{B}}$ un sous groupe de Borel de $\underline{\underline{G}}$ contenant $\underline{\underline{H}}$, et $\underline{\underline{U}}$ le radical unipotent de $\underline{\underline{B}}$. On notera $\underline{\underline{\mathfrak{g}}}$, $\underline{\underline{\mathfrak{h}}}$, $\underline{\underline{\mathfrak{b}}}$, $\underline{\underline{\mathfrak{u}}}$ les algèbres de Lie de $\underline{\underline{G}}$, $\underline{\underline{H}}$, $\underline{\underline{B}}$, $\underline{\underline{U}}$ respectivement, et G , H , B , U les ensembles des points de $\underline{\underline{G}}$, $\underline{\underline{H}}$, $\underline{\underline{B}}$, $\underline{\underline{U}}$ rationnels sur K. Chacun est canoniquement muni d'une topologie qui en fait un groupe localement compact totalement discontinu. De même, on notera $\mathfrak{g}$, $\mathfrak{b}$, $\mathfrak{h}$, $\mathfrak{u}$ les ensembles des points de $\underline{\underline{\mathfrak{g}}}$, $\underline{\underline{\mathfrak{b}}}$, $\underline{\underline{\mathfrak{h}}}$, $\underline{\underline{\mathfrak{u}}}$ rationnels sur K. Ce sont des algèbres de Lie sur K.

Si X est un groupe localement compact, on notera $d_X\, x$ (ou dx s'il n'y a pas d'ambiguité) une mesure de Haar à droite sur X.

II.2. Nous allons étudier des représentations du groupe G. Il faut avant cela en préciser la structure. Soit R le système des racines de $\mathfrak{g}$ par rapport à $\mathfrak{h}$. Si α est une racine, on note $\mathfrak{g}^{\alpha}$ la sous algèbre correspondante de $\underline{\mathfrak{g}}$, formée des éléments x tels que

$$[h,x] = \alpha(h)\, x$$

si h est un élément de $\mathfrak{h}$. Les racines positives seront celles telles que $\mathfrak{g}^\alpha$ soit contenu dans $\mathfrak{u}$, et on notera S le système de racines simples correspondant. Pour chaque racine α, on choisit une base X_α de $\mathfrak{g}^\alpha$ qui soit définie sur K, et telle que le système des X_α forme une base de Chevalley de $\mathfrak{g}$ ([8] et [18] Théorème 1 p.6). On définit de même un sous-groupe $\underline{\underline{U}}^\alpha$ de $\underline{\underline{G}}$ et un homomorphisme de groupes algébriques x_α de $\bar{K}$ sur $\underline{\underline{U}}^\alpha$ tel que

$$h\, x_\alpha(t)\, h^{-1} = x_\alpha(\alpha(h)t)$$

si h est dans $\underline{\underline{H}}$ et t dans $\bar{K}$. On fixe x_α par la condition que le vecteur tangent à x_α à l'origine est X_α. On note U^α l'intersection de $\underline{\underline{U}}^\alpha$ avec G. La restriction de x_α à K est un homomorphisme, encore noté x_α de K sur U^α. Le système $(H,(U^\alpha)_{\alpha\in R})$ est alors une donnée radicielle génératrice de type R dans G, au sens de [5] (6.1). On définit, pour chaque racine α, une application φ_α de U^α dans $\mathbb{Z} \cup \{\infty\}$ par la relation

$$\varphi_\alpha \circ x_\alpha = v_K .$$

On définit de même une application φ_o de H dans $\mathbb{Z} \cup \{\infty\}$ par la relation :

$$\varphi_o(h) = \operatorname*{Inf}_{\chi\in X(\underline{\underline{H}})} v_K(\chi(h) - 1)$$

où h est dans H et $X(\underline{\underline{H}})$ est l'ensemble des caractères rationnels de $\underline{\underline{H}}$ (i.e. des homomorphismes de groupes algébriques de $\underline{\underline{H}}$ dans le groupe multiplicatif $\bar{K}^\times$ de $\bar{K}$). Le système $((\varphi_\alpha)_{\alpha\in R}, \varphi_o)$ est alors une valuation prolongée de la donnée radicielle $(H,(U^\alpha)_{\alpha\in R})$([5], définition 6.4.38). Si n est un entier, on définit des sous-groupes U^α_n et H_n de G par les relations suivantes :

$$U^\alpha_n = \varphi_\alpha^{-1}([n\, \infty]) = x_\alpha(p^n)$$

$$H_n = \varphi_o^{-1}([n\, \infty]) = \{h \in H | \chi(h) \in 1 + p^n \text{ pour tout } \chi \text{ dans } X(\underline{\underline{H}})\}$$

Soit G_n le sous-groupe de G engendré par les sous-groupes H_n et U^α_n pour chaque racine α. Si X est une partie quelconque de G, on note $X_n = X \cap G_n$. On note $\bar{U}$ le sous groupe de G engendré par les sous groupes U^α où α est une racine négative.

<u>Lemme 1</u>

(i) Si $n \geqslant 1$, alors $G_n = \bar{U}_n H_n U_n$,

(ii) $U_n = \prod_{\alpha>0} U_{\alpha,n}$,

les racines α étant prises dans un ordre quelconque.

(iii) Si $n \geqslant 1$, alors G_n/G_{2n} est un groupe commutatif.

(iv) $\bigcap_{n \geqslant 1} G_n = \{e\}$.

Démonstration

(i) découle de [5], (proposition 6.4.48) ,

(ii) de [5] (proposition 6.4.9 (ii))

(iii) est une conséquence de [5], (proposition 6.4.44), enfin (iv) est une conséquence de (i) et (ii). Ce lemme est aussi démontré dans le cas des groupes de Chevalley dans [9] (lemme 4).

II.3. Rappels sur les représentations

Nous adoptons ici les définitions de [10].

Si Γ est un groupe localement compact totalement discontinu, on appellera représentation de Γ dans un espace vectoriel complexe E un homomorphisme de Γ dans le groupe des automorphismes de E. Une représentation de Γ est dite de classe C^∞ si le stabilisateur de tout point de E est un sous-groupe ouvert de Γ . Elle est dite admissible si, de plus, l'ensemble des éléments de E invariants par un sous-groupe ouvert de Γ est de dimension finie. Si Γ_1 est un sous-groupe fermé de Γ et π une représentation de Γ_1 dans un espace E, on note $C_c^\infty(\Gamma,E,\pi)$ l'espace des fonctions f de Γ dans E telles que :

- $f(g_1\, g) = \pi(g_1)\, f(g)$ si $g_1 \in \Gamma_1$ et $g \in \Gamma$,

- f est localement constante, et son support est compact modulo Γ_1.

Le groupe Γ opère par translations à droite dans $C_c^\infty(\Gamma,E,\pi)$ et la représentation de Γ ainsi définie est la représentation induite par π , que l'on notera $\mathrm{Ind}_{\Gamma_1}^{\Gamma}\,\pi$. Soit $C_c^\infty(\Gamma;E)$ l'espace des fonctions de Γ dans E localement constantes à support compact. L'application Δ de $C_c^\infty(\Gamma;E)$ dans $C_c^\infty(\Gamma,E,\pi)$ donnée par :

$$\Delta(f)(g) = \int_{\Gamma_1} \pi(g_1)^{-1}\, f(g_1\, g)\, d_{\Gamma_1}\, g_1$$

est un homomorphisme de $C_c^\infty(\Gamma,E)$ sur $C_c^\infty(\Gamma,E,\pi)$ commutant avec l'action de Γ par translations à droite ([16], I,2, Proposition 1).

Si π_1 et π_2 sont deux représentations de Γ , on note $\mathrm{Hom}_\Gamma(\pi_1,\pi_2)$ l'espace des opérateurs d'entrelacement de π_1 dans π_2.

Si π est une représentation admissible de Γ dans un espace E, et si f est un élément de $C_c^{\infty}(G)$, alors l'opérateur $\pi(f)$ défini par

$$\pi(f)\, x = \int_{\Gamma} f(\gamma)\ \pi(\gamma)x \quad d\gamma$$

si x est un élément de E, est de rang fini. On appelle caractère de π la forme linéaire Θ_π sur $C_c^{\infty}(G)$ définie par :

$$\Theta_\pi(f) = \mathrm{Tr}\ \pi(f)$$

III - Un système inductif de représentations.

Nous énonçons ici le résultat essentiel de ce travail : la construction d'un système inductif de représentations de G et la détermination de sa limite.

Nous allons définir des représentations de G comme représentations induites. Soit τ un caractère du groupe additif de K dont le noyau soit égal à $\mathcal{O}$. D'après le lemme 1 la formule suivante définit pour tout entier $n \geq 1$ une caractère (i.e. une représentation de dimension 1), noté Ψ_n , de G_n :

$$\Psi_n(bu) = \prod_{\alpha \in S} \tau(t_\alpha\, \varpi^{-2n})$$

où b est dans $\bar{B}_n$ $(= H_n \bar{U}_n)$ et u est un élément de U_n donné par $u = \prod_{\alpha>0} x_\alpha(t_\alpha)$.

Soit d un élément de H tel que $\alpha(d) = \varpi^{-2}$ pour toute racine simple α (il est aisé de voir qu'un tel élément existe toujours). Nous transformons G_n et Ψ_n par automorphisme intérieur en posant, pour tout entier $n \geq 1$,

$$V_n = d^n\, G_n\, d^{-n}$$

$$\theta_n(u) = \Psi_n(d^{-n}\, u\, d^n) \quad \text{si} \quad u \in V_n .$$

On définit un caractère θ de U par

$$\theta(u) = \prod_{\alpha \in S} \tau(t_\alpha)$$

si $$u = \prod_{\alpha>0} x_\alpha(t_\alpha) .$$

Lemme 2

i) La suite $V_n \cap U$ est une suite croissante de sous-groupes compacts ouverts de U dont la réunion est égale à U.

ii) La suite $V_n \cap \bar{B}$ est une suite décroissante de sous-groupes compacts ouverts de $\bar{B}$ dont l'intersection est réduite à l'élément neutre.

iii) Les caractères θ_n et θ coïncident sur $V_n \cap U$.

Démonstration

Le lemme 2 découle trivialement du lemme 1 et des définitions.

On peut maintenant définir pour chaque entier $n \geq 1$ une représentation κ_n de G

$$\kappa_n = \mathrm{Ind}_{V_n}^{G} \theta_n$$

On note K_n l'espace de la représentation κ_n. Si m et n sont deux entiers supérieurs ou égaux à 1, la formule suivante définit un opérateur d'entrelacement A_n^m de K_n dans K_m :

$$A_n^m f(g) = \frac{1}{\mathrm{vol}\ (V_m)} \int_{V_m} f(ug)\ \overline{\theta_m(u)}\ du$$

où f est dans K_n et g dans G ; Vol (V_m) désigne le volume de V_m.

Lemme 3

Si $m \geq n \geq 1$, on a

$$A_n^m f(g) = \frac{1}{\mathrm{vol}\ (V_m \cap U)} \int_{V_m \cap U} f(ug)\ \overline{\theta(u)}\ d_{V_m \cap U} u$$

Démonstration

C'est clair.

Proposition 1

Le système $((\kappa_n)_{n \geq 1}\ ,\ (A_n^m)_{m \geq n \geq 1})$ est un système inductif de représentations de G.

Démonstration

Il suffit de montrer que, si $l \geq m \geq n \geq 1$, alors on a : $A_m^l \circ A_n^m = A_n^l$, ce qui est immédiat d'après les lemmes 2 et 3.

On notera $\mathcal{K}$ ce système inductif.

Proposition 2

Le système inductif $\mathcal{K}$ admet une limite.

Démonstration

C'est clair.

Notons $(\kappa_\infty, (A_n)_{1 \leq n})$ cette limite. Nous allons construire une représentation équivalente à κ_∞. Soit la représentation induite $\lambda_\theta = \mathrm{Ind}_U^G \theta$, dans l'espace Λ_θ. On définit un opérateur d'entrelacement ϕ_n de κ_n dans λ_θ par

$$\phi_n f(g) = \int_U f(ug)\,\overline{\theta(u)}\,du$$

où f est dans K_n et g dans G. D'après le lemme 3, si $m \geqslant n \geqslant 1$, on a $\phi_m \circ A_n^m = \phi_n$, donc il existe un opérateur d'entrelacement unique ϕ_∞ de K_∞ dans Λ_θ tel que $\phi_\infty \circ A_n = \phi_n$ pour tout $n \geqslant 1$.

Proposition 3

ϕ_∞ est un isomorphisme de la représentation κ_∞ avec λ_θ.

Démonstration

Pour prouver la surjectivité de ϕ_∞, il suffit de montrer que, si f est un élément de Λ_θ, alors il existe un nombre entier n et un élément φ de K_n tel que $\phi_n(\varphi) = f$. En effet un élément f de Λ_θ est donné par

$$f(g) = \int_U \varphi_1(ug)\,\overline{\theta(u)}\,du$$

où φ_1 est un élément de $C_c^\infty(G)$. Soit n un entier tel que φ_1 soit invariante par les translations à gauche par les éléments de G_n, et soit φ l'élément de K_n défini par :

$$\varphi(g) = \frac{1}{\mathrm{vol}\,(V_n)} \int_{V_n} \varphi_1(ug)\,\overline{\theta_n(u)}\,du$$

Comme φ_1 est invariante à gauche par $V_n \cap \overline{B}$, on a

$$\varphi(g) = \frac{1}{\mathrm{vol}(V_n \cap U)} \int_{V_n \cap U} \varphi_1(ug)\,\overline{\theta(u)}\,du$$

On peut donc calculer $\phi_n(\varphi)$:

$$\begin{aligned}\phi_n(\varphi)(g) &= \frac{1}{\mathrm{vol}(V_n \cap U)} \int_U \int_{V_n \cap U} \varphi_1(u_1\ u\ g)\,\overline{\theta(u)}\,\overline{\theta(u_1)}\,du\,du_1 \\ &= \int_U \varphi_1(ug)\,\overline{\theta(u)}\,du \\ &= f(g)\end{aligned}$$

Donc φ est bien l'élément cherché de K_n.

Montrons maintenant l'injectivité de ϕ_∞. Il suffit pour cela de prouver que, si n est un entier et f un élément de K_n tel que $\phi_n(f) = 0$, alors il existe un entier $m \geq n$ tel que $A_n^m f = 0$ (et par conséquent $A_m(f) = 0$). En effet, soit un entier $m \geq n$ et f un élément de K_n. Alors, d'après le lemme 3, le support de $A_n^m f$ est contenu dans $(V_m \cap U)$ Supp f. De plus, si u est un élément de $V_m \cap U$ et g un élément de g , on a

$$A_n^m f(ug) = \theta(u) A_m^n f(g) .$$

Par conséquent $A_n^m f$ est nul dès que sa restriction au support de f est nulle. D'autre part, comme $V_m \cap U$ est un sous-groupe ouvert de U, on peut choisir comme mesure de Haar sur $V_m \cap U$ la restriction de la mesure de Haar de U. On a alors, si $\phi_m(f) = 0$,

$$A_n^m f(g) = \frac{1}{\mathrm{vol}(V_m \cap U)} \left[\int_{V_m \cap U} f(ug)\, \overline{\theta(u)}\, du - \int_U f(ug)\, \overline{\theta(u)}\, du \right]$$

$$= \frac{-1}{\mathrm{vol}(V_m \cap U)} \int_{U-(V_m \cap U)} f(ug)\, \overline{\theta(u)}\, du \quad .$$

D'après le lemme 2 (i), et comme le support de f est compact, on peut choisir l'entier $m \geq n$ tel que

$$V_m \cap U \supset (\mathrm{Supp}\ f)\ (\mathrm{Supp}\ f)^{-1} \cap U$$

c'est-à-dire tel que

$$(U - (V_m \cap U)) \cap (\mathrm{Supp}\ f)(\mathrm{Supp}\ f)^{-1} = \emptyset .$$

Si g est un élément de Supp f et u un élément de $U - (V_m \cap U)$, on a donc :

$$ug \notin \mathrm{Supp}\ f$$

Par conséquent, si g est un élément de Supp f , on a :

$$\int_{U-(V_m \cap U)} f(ug)\, \overline{\theta(u)}\, du = 0$$

Donc $A_n^m f = 0$.

IV - Propriétés du système inductif.

Soit α_0 la plus grande racine de R, et μ sa hauteur : si $\alpha_0 = \sum_{\alpha \in S} n_\alpha \alpha$ est l'expression de α_0 dans la base S, on a $\mu = \sum_{\alpha \in S} n_\alpha$.

A partir de maintenant, nous supposons que la caractéristique résiduelle p de

K vérifie $p \geqslant 2\mu + 1$, et que la caractéristique de K est nulle.

Nous fixons la mesure de Haar dg sur G. L'espace $C_c^\infty(G)$ est alors une algèbre pour le produit de convolution, noté $*$. Nous choisirons comme mesure de Haar sur chaque sous groupe ouvert V_n la restriction de dg à V_n. Nous notons $\varepsilon(g)$ la mesure de Dirac en un point g de G. Nous identifions un élément de $C_c^\infty(V_n)$ à un élément de $C_c^\infty(G)$ en le prolongeant par 0 à l'extérieur de V_n.

Lemme 4

Il existe un entier n_o tel que, si n est un entier au moins égal à n_o, et u un élément de U tel que

$$\theta_n * \varepsilon(u) * \theta_n \neq 0$$

on ait $u \in V_n$.

La démonstration de ce lemme fera l'objet du paragraphe suivant.

Lemme 5

Si $m \geqslant n \geqslant n_o$, on a :

$$\theta_n * \theta_m * \theta_n = \mathrm{vol}\,(V_n)\ \mathrm{vol}\,(V_m \cap V_n)\ \theta_n$$

Démonstration

D'après le lemme 2, il existe un nombre fini d'éléments u_i de $U \cap V_m$ qui forment un système de représentants de $V_m/(V_n \cap V_m)$. Si $1_{V_m \cap V_n}$ désigne la fonction caractéristique de $V_m \cap V_n$, on a

$$\theta_n * \theta_m * \theta_n = \sum_i \theta_n * \varepsilon(u_i) * \theta_n\ 1_{V_m \cap V_n} * \theta_n$$

$$\theta_n * \theta_m * \theta_n = \sum_i \theta_m(u_i)\ \theta_n * \varepsilon(u_i) * \theta_n\ 1_{V_m \cap V_n} * \theta_n$$

$$\theta_n * \theta_m * \theta_n = \sum_i \theta_m(u_i)\ \mathrm{vol}\,(V_m \cap V_n)\ \theta_n * \varepsilon(u_i) * \theta_n$$

$$= \mathrm{vol}\,(V_m \cap V_n)\ \theta_n * \theta_n\ ,$$

d'après le lemme 4, d'où la conclusion.

Proposition 4

Si $p \geqslant 2\mu + 1$ et si $m \geqslant n \geqslant 1$, on a $A_m^n A_n^m = \frac{\mathrm{vol}(V_m \cap V_n)}{\mathrm{vol}\, V_m} \mathrm{Id}_{K_n}$. En particulier A_n^m est injectif et A_m^n est surjectif.

Démonstration

Remarquons que K_n est l'idéal à droite de $C_c^\infty(G)$ engendré par θ_n, et que l'opérateur A_n^m n'est autre que la convolution à gauche par la fonction $\frac{1}{\mathrm{vol}\, V_m}\,\theta$. Il suffit donc de calculer $A_m^n\, A_n^m\, \theta_n$ qui est égal, d'après ce qui précède, à $\frac{1}{\mathrm{vol}\, V_m\, \mathrm{vol}\, V_n}\,\theta_n * \theta_m * \theta_n$. On a donc, d'après le lemme 5

$$A_m^n\, A_n^m\, \theta_n = \frac{\mathrm{vol}(V_m \cap V_n)}{\mathrm{vol}\, V_m}\,\theta_n$$

Corollaire 1

Les opérateurs A_n et ϕ_n sont injectifs si $n \geqslant n_o$.

Démonstration

C'est clair pour les opérateurs A_n qui s'obtiennent par passage à la limite, et pour les opérateurs ϕ_n car $\phi_n = \phi_\infty \circ A_n$.

Corollaire 2

Soit π une représentation de G. Les applications

$$\mathrm{Hom}_G(K_m, \pi) \longrightarrow \mathrm{Hom}_G(K_n, \pi)$$
$$f \longmapsto f \circ A_n^m$$

et

$$\mathrm{Hom}_G(\lambda_\theta, \pi) \longrightarrow \mathrm{Hom}_G(K_n, \pi)$$
$$f \longmapsto f \circ \phi_n$$

sont surjectives si $m \geqslant n \geqslant n_o$.

Démonstration

C'est immédiat par dualité à partir de la Proposition 4 et du Corollaire 1.

Corollaire 3

Soit π une représentation de G telle que

$$\sup_n \dim \mathrm{Hom}_G(K_n, \pi) < +\infty$$

Alors, on a

$$\dim \mathrm{Hom}_G(\lambda_\theta, \pi) = \lim_n \dim \mathrm{Hom}_G(K_n, \pi)$$

Démonstration

Il suffit de voir que $\mathrm{Hom}_G(\lambda_\theta, \pi)$ est la limite projective du système des espaces vectoriels $\mathrm{Hom}_G(K_n, \pi)$ avec les applications surjectives définies dans le Corollaire 2 : cela résulte, par dualité, de la proposition 3 ([1], ch. II, §6, Prop. 6). Le corollaire est alors immédiat, car cette suite est stationnaire d'après l'hypothèse.

Lemme 6

Si π est une représentation admissible de G dans un espace E, alors

$$\dim \mathrm{Hom}_G (K_n, \pi) = \frac{1}{\mathrm{vol}\,(G_n)} \quad \Theta_\pi(\Psi_n)$$

Démonstration

Par réciprocité de Frobenius ([16], corollaire 2 du théorème 1), on a

$$\dim \mathrm{Hom}_G (K_n, \pi) = \dim \mathrm{Hom}_{G_n} (\Psi_n , \pi|_{G_n})$$

car K_n est isomorphe à $\mathrm{Ind}_{G_n}^G \Psi_n$. L'espace $\mathrm{Hom}_{G_n} (\Psi_n , \pi|_{G_n})$ est isomorphe au sous-espace E_1 de E formé des éléments x tels que $\pi(g)x = \Psi_n(g)x$ si $g \in G_n$. D'après le théorème de Peter-Weyl, l'opérateur $\frac{1}{\mathrm{vol}\, G_n} \pi(\Psi_n)$ est un projecteur de E sur E_1 et sa trace est donc égale à la dimension de E_1, d'où le lemme.

Théorème

Supposons que la caractéristique de K soit nulle et que $p \geq 2\mu + 1$. Soit π une représentation admissible de G, de caractère Θ_π. Si $\sup_n [\frac{1}{\mathrm{vol}\,(G_n)} \Theta_\pi(\Psi_n)] < +\infty$ alors

$$\dim \mathrm{Hom}_G (\lambda_\theta, \pi) = \lim_n \left[\frac{1}{\mathrm{vol}\,(G_n)} \Theta_\pi(\Psi_n)\right].$$

Démonstration

C'est clair d'après le lemme 6, et le corollaire 3 de la proposition 4.

Comme application, nous avons par exemple le

Corollaire 1

Supposons que $\Theta_\pi = \sum \lambda_i \Theta_{\pi_i}$, où π est une représentation admissible de G, et les π_i sont des représentations admissibles de G en nombre fini, vérifiant les conditions du théorème, et les λ_i des entiers. Alors

$$\dim \mathrm{Hom}_G(\lambda_\theta, \pi) = \sum \lambda_i \dim \mathrm{Hom}_G (\lambda_\theta , \pi_i)$$

Ces résultats permettent de préciser la restriction d'une représentation au sous groupe G_o de G.

Corollaire 2

Soit π une représentation admissible irréductible de G, et soit $\mathfrak{D}$ une composante irréductible de la représentation $\mathrm{Ind}_{G_n}^{G_o} \psi_n$ de G_o, où $n \geq n_o$. Alors

$$\dim \mathrm{Hom}_{G_o} (\mathfrak{D}, \pi|_{G_o}) \leq 1 .$$

Démonstration

On a

$$\dim \mathrm{Hom}_{G_o} (\mathfrak{D}, \pi|_{G_o}) \leq \dim \mathrm{Hom}_{G_o} (\mathrm{Ind}_{G_n}^{G_o} \psi_n , \pi|_{G_o}) .$$

Par réciprocité de Frobenius, ([16], Cor. 2 du Th. 2), on a

$$\dim \mathrm{Hom}_{G_o} (\mathrm{Ind}_{G_n}^{G_o} \psi_n , \pi|_{G_o}) = \dim \mathrm{Hom}_G (\mathrm{Ind}_{G_n}^{G} \psi_n , \pi) .$$

Le corollaire 2 de la proposition 4 nous donne

$$\dim \mathrm{Hom}_G (\kappa_n , \pi) \leq \dim \mathrm{Hom}_G (\lambda_\theta , \pi)$$

Enfin on a, d'après le théorème 2 de [16] (voir aussi [15], théorème 3),

$$\dim \mathrm{Hom}_G (\lambda_\theta , \pi) \leq 1$$

D'où la conclusion.

Remarque 1

Comme conséquence du corollaire 2, on peut montrer par exemple que la représentation de Steinberg de G ([10], § 15 ; [6]) admet un modèle de Whittaker ([15]). Cependant une méthode due à Casselman permet de montrer ce résultat sans les restrictions sur le corps K que nous avons faites ici (cf. [6] et [7]).

Remarque 2

Soit π une représentation admissible irréductible parabolique (i.e. "supercuspidale") de G, et Θ_π son caractère. Harish-Chandra et R. Howe ont montré (non publié, mais cf. [11]) qu'il existe un voisinage $\mathcal{V}$ de 0 dans $\mathfrak{g}$ tel que, si Ξ est la distribution sur $\mathcal{V}$ définie par

$$\Xi(f) = \Theta_\pi(f \circ \log)$$

où log est l'application logarithme de G, on ait

$$\Xi = \sum_x c(x)\ \theta_x$$

où la sommation est étendue aux classes de conjugaison d'éléments nilpotents de $\mathfrak{g}$, les c(x) sont des entiers, et θ_x est la transformée de Fourier sur $\mathfrak{g}$ d'une mesure invariante sur la classe x.

Alors, d'après le théorème, on peut associer au caractère θ de G une classe x_o d'éléments nilpotents réguliers telle que

$$c(x_o) = \dim \operatorname{Hom}_G (\lambda_\theta , \pi)$$

à condition que θ_{x_o} soit convenablement normalisé.

V - <u>Démonstration du lemme 4</u>

V.1. Nous reprenons les mêmes hypothèses sur K : la caractéristique de K est nulle, et $p \geqslant 2\mu +$

Le groupe G est muni d'une structure de variété analytique sur K, qui en fait un groupe de Lie sur K, d'algèbre de Lie $\mathfrak{g}$ (cf. [19], appendice 3).

On va définir une suite de voisinages de l'origine dans $\mathfrak{g}$. Si n est un entier soit $\mathfrak{h}_n$ l'ensemble des éléments h de $\mathfrak{h}$ tels que $\langle d\chi, h\rangle \in \mathfrak{p}^n$ pour tout élément χ de $X(\underline{\underline{H}})$, dont la forme linéaire tangente est notée $d\chi$. Soit $\mathfrak{g}_n$ le sous $\mathcal{O}$-module de $\mathfrak{g}$ engendré par $\mathfrak{h}_n$ et les $\mathfrak{p}^n X_\alpha$. Alors $[\mathfrak{g}_n , \mathfrak{g}_n] \subset \mathfrak{g}_n$ et $\mathfrak{g}_n = \varpi^n \mathfrak{g}_o$. Si X est une partie de $\mathfrak{g}$, on note $X_n = X \cap \mathfrak{g}_n$.

D'après [3], (chapitre III, §7, n°2, Prop. 3), il existe un entier ν et une application exponentielle de G, notée exp, définie sur $\mathfrak{g}_\nu$, et dont l'image exp $\mathfrak{g}_\nu$ est un sous groupe de G. On a $\exp (t X_\alpha) = x_\alpha (t)$ et, si $\chi \in X(\underline{\underline{H}})$, et $h \in \mathfrak{h}$, on a $\chi(\exp h) = \exp(\langle d\chi , h\rangle)$. On en déduit que $\exp (\mathfrak{g}_\nu) = G_\nu$.

<u>Lemme 7</u>

Si n est assez grand, la composée de exp avec l'application canonique de G_n dans G_n/G_{2n} est un homomorphisme de groupes de $\mathfrak{g}_n$ dans G_n/G_{2n}.

<u>Démonstration</u>

La série de Hausdorff de $\mathfrak{g}$ ([3], chapitre II, §6, n°4) est convergente sur $\mathfrak{g}_n$ si n est assez grand et définit une loi de composition $(x,y) \longmapsto H(x,y)$ sur $\mathfrak{g}_n$ qui en fait un groupe de Lie sur K ; l'application exponentielle est, pour n assez grand, un homomorphisme de groupes de $\mathfrak{g}_n$ dans G_n (loc. cit.). Le calcul des termes de la série de Hausdorff montre que, si x et y sont des éléments de $\mathfrak{g}_n$,

$$(*) \qquad H(x,y) = x + y \pmod{\mathfrak{g}_{2n}}$$

On a $H(x,y) \in \mathfrak{g}_n$. On a donc, en appliquant à nouveau cette congruence :

$$H(H(x,y)\ ,\ -(x+y)) \equiv H(x,y) - (x+y) \pmod{\mathfrak{g}_{2n}}$$

D'où, en utilisant (*)

$$H(H(x,y)\ ,\ -(x+y)) \in \mathfrak{g}_{2n}$$

Par conséquent

$$\exp x \exp y\ (\exp (x+y))^{-1} = \exp [H(H(x,y)\ ,\ -(x+y))] \in G_{2n}$$

D'où le résultat.

V.2. D'après le lemme 7, $\Psi_n \circ \exp$ est un caractère du groupe additif de $\mathfrak{g}_n$, pour n assez grand. Soit $\tilde{\Psi}_n = \Psi_n \circ \exp$. Soit $\tilde{\Psi}$ le caractère de $\mathfrak{g}$ défini par

$$\tilde{\Psi}(t\, X_\alpha) = \Upsilon(t) \text{ si } \alpha \in S \text{ et } t \in K.$$
$$\tilde{\Psi}(x) = 1 \quad \text{si} \quad x \in \mathfrak{h} + \sum_{\alpha \notin S} \mathfrak{g}^\alpha .$$

Lemme 8

Si x est un élément de $\mathfrak{g}_n$, alors

$$\tilde{\Psi}_n(x) = \tilde{\Psi}(\varpi^{-2n}\, x)$$

Démonstration

Il suffit de le vérifier pour $x = t\, X_\alpha$, ou $x \in \mathfrak{h}$, auxquels cas c'est clair.

Notons $\hat{\mathfrak{g}}$ l'ensemble des caractères de $\mathfrak{g}$. Les opérateurs sur $\mathfrak{g}$ définissent par transposition des opérateurs sur $\hat{\mathfrak{g}}$, d'où la structure d'espace vectoriel de $\hat{\mathfrak{g}}$ et les représentations Ad^*(resp. ad^*) de G (resp. $\mathfrak{g}$) dans $\hat{\mathfrak{g}}$ telles que :

$$Ad^*\, g = {}^t(Ad\, g^{-1})$$

$$ad^*\, x = -\, {}^t(ad\, x)$$

Si X est une partie de $\hat{\mathfrak{g}}$ et n un entier, on note $X_n = \mathfrak{g}^{\perp}_{-n} \cap X$.

Lemme 9

Si $\mathfrak{a}$ est un sous espace vectoriel de $\hat{\mathfrak{g}}$ tel que $\mathfrak{a}_o + (ad^*\, \mathfrak{g}_o)\tilde{\Psi} = \mathfrak{g}_o^{\perp}$, alors , si n est assez grand, l'ensemble des caractères de $\mathfrak{g}$ dont la restriction à $\mathfrak{g}_n$ coïncide avec $\tilde{\Psi}_n$ est égal à

$$\mathrm{Ad}^* G_n (\varpi^{-2n}\tilde{\Psi} + \mathfrak{a}_{-n})$$

Démonstration

Somme Ψ_n est invariant par G_n, l'ensemble des caractères de $\mathfrak{g}$ dont la restriction à $\mathfrak{g}_n$ coïncide avec $\tilde{\Psi}_n$ est invariant par G_n, et il est clair qu'il est égal à $\varpi^{-2n}\tilde{\Psi} + \mathfrak{g}_n^{\perp}$. Il contient donc $\mathrm{Ad}^* G_n (\varpi^{-2n}\tilde{\Psi} + \mathfrak{a}_{-n})$. Reste à montrer l'inclusion

$$\mathrm{Ad}^* G_n (\varpi^{-2n}\tilde{\Psi} + \mathfrak{a}_{-n}) \supset \varpi^{-2n}\tilde{\Psi} + \mathfrak{g}_n^{\perp}$$

Il suffit pour cela de montrer que, si χ est un élément de $\mathfrak{g}_n^{\perp}$, l'application P de $\mathfrak{g}_o$ dans $\mathfrak{g}_o^{\perp}/\mathfrak{a}_o$ donnée par

$$P(y) = \varpi^n \left[(\mathrm{Ad}^*(\exp \varpi^n y))(\varpi^{-2n}\tilde{\Psi} + \chi) - \varpi^{-2n}\tilde{\Psi} + \mathfrak{a}_{-n} \right]$$

prend la valeur 0. Or P admet le développement en série :

$$P(y) = \varpi^n \chi + \sum_{m \geq 1} \frac{\varpi^{n(m-1)}}{m!} (\mathrm{ad}^* y)^m (\tilde{\Psi} + \varpi^{2n}\chi) + \mathfrak{a}_o$$

Avec les hypothèses faites sur $\mathfrak{a}$, le lemme de Hensel ([2], Chapitre III, §4, n°5, Cor. 2 du Th. 2) permet d'affirmer que P admet un zéro.

Lemme 10

Il existe un isomorphisme de $\mathfrak{g}$-modules η de $\mathfrak{g}$ dans $\hat{\mathfrak{g}}$ qui envoie $\mathfrak{g}_o$ dans $\hat{\mathfrak{g}}_o$.

Démonstration.

Soit $\underline{\underline{Z}}$ le centre de $\underline{\underline{G}}$, $\underline{\underline{G}}'$ le sous groupe dérivé et $\underline{\underline{H}}' = \underline{\underline{H}} \cap \underline{\underline{G}}'$. On a la suite exacte

$$0 \longrightarrow \underline{\underline{H}}' \cap \underline{\underline{Z}} \longrightarrow \underline{\underline{H}}' \times \underline{\underline{Z}} \longrightarrow \underline{\underline{H}} \longrightarrow 0$$

Par tranposition on obtient la suite exacte

$$0 \longrightarrow X(\underline{\underline{H}}) \longrightarrow X(\underline{\underline{H}}') \times X(\underline{\underline{Z}}) \longrightarrow X(\underline{\underline{H}}' \cap \underline{\underline{Z}}) \longrightarrow 0$$

Soit L_p le réseau des poids du système de racines de $\underline{\underline{G}}'$ par rapport à $\underline{\underline{H}}'$, et L_R le réseau des poids radiciels. On a la suite exacte

$$0 \longrightarrow L_R \longrightarrow X(\underline{\underline{H}}') \longrightarrow X(\underline{\underline{H}}' \cap \underline{\underline{Z}}) \longrightarrow 0$$

Donc les groupes $X(\underline{\underline{H}}')/L_R$ et $(X(\underline{\underline{H}}') \times X(\underline{\underline{Z}}))/X(\underline{\underline{H}})$ sont isomorphes. Par conséquent, les groupes $(X(\underline{\underline{Z}}) X L_p)/X(\underline{\underline{H}})$ et L_P/L_R ont même ordre, lequel est l'indice de connexion du système de racine R, et qui, comme on le vérifie facilement, est inversible dans $\mathcal{O}$. Par conséquent, si nous identifions $X(\underline{\underline{Z}}) \times L_P$ à un sous groupe du dual $\mathfrak{h}^*$ de $\mathfrak{h}$, on a

$$\mathfrak{h}_o = \{h \in \mathfrak{h} \mid \gamma(\mathfrak{h}_o) \in \mathcal{O} \text{ pour tout } \gamma \in X(\underline{\underline{Z}}) \times L_P\}$$

Soit $\mathfrak{z}$ le centre de $\mathfrak{g}$, $\mathfrak{g}'$ l'algèbre dérivée de $\mathfrak{g}$. On a $\mathfrak{g} = \mathfrak{g}' \oplus \mathfrak{z}$, et, d'après ce qui précède, $\mathfrak{g}_o = (\mathfrak{g}' \cap \mathfrak{g}_o) \oplus (\mathfrak{z} \cap \mathfrak{g}_o)$. Pour définir η, il suffit de le définir sur $\mathfrak{z}$, ce qui est évident, et sur $\mathfrak{g}'$, ce qui résulte du lemme 6.1 de [4].

<u>Remarque</u>

Soit X_α^* l'élément de $\hat{\mathfrak{g}}$ tel que

$$X_\alpha^*(x) = \begin{cases} 1 & \text{si } x \in \mathfrak{h} + \sum_{\alpha \neq \beta} \mathfrak{g}^\beta \\ \tau(t) & \text{si } x = tX_\alpha \end{cases}$$

Avec la construction utilisée, on a $\eta(X_\alpha) = C_\alpha X_{-\alpha}^*$ où $C_\alpha \in \mathcal{O}^*$.

Nous allons maintenant choisir le sous espace $\mathfrak{a}$ qui intervient dans le lemme 9. Soit $e_- = \eta^{-1}(\check{\Psi})$. D'après la remarque précédente, on a

$$e_- = \sum_{\alpha \in S} C_\alpha^{-1} X_{-\alpha}$$

avec $C_\alpha^{-1} \in \mathcal{O}^*$. Pour chaque $\alpha \in S$, soit $H_\alpha = [X_\alpha, X_{-\alpha}]$. Les H_α forment le système de racines inverse de R ; soit h_o l'élément de $\mathfrak{h}$ qui est le double de la somme des poids fondamentaux de ce système. On a donc $\alpha(h_o) = 2$ pour tout $\alpha \in S$, et $h_o = \sum_{\alpha \in S} r_\alpha H_\alpha$ où les r_α sont des entiers. Posons

$$e_+ = \sum_{\alpha \in S} C_{-\alpha} r_\alpha X_\alpha$$

Alors on définit $\mathfrak{a}$ comme l'image par η du centralisateur de e_+ dans $\mathfrak{g}$.

<u>Lemme 11</u>

i) $\mathfrak{a}_o \oplus (\mathrm{ad}^* \mathfrak{g}_o)\check{\Psi} = \mathfrak{g}_o^\perp$

ii) Si deux éléments de $\check{\Psi} + \mathfrak{a}$ sont dans la même orbite de G (pour son action dans $\hat{\mathfrak{g}}$ par Ad^*), alors ils sont égaux.

Démonstration

Les relations de commutation entre e_+ , h_o, et e_- montrent que ces trois éléments engendrent sur $\mathcal{O}$ une algèbre de Lie $\mathfrak{s}$ isomorphe à $\mathfrak{sl}_2(\mathcal{O})$. D'après [4] (lemme 1.2), la représentation adjointe de s dans $\mathfrak{g}_o$ est complètement réductible , et dans chaque composante irréductible, l'image de ad e_- admet le centralisateur de e_+ comme supplémentaire. Cela montre i). L'assertion ii) se déduit de [14] (théorème 0.10).

V.3. Rappelons qu'un élément x de $\mathfrak{g}$ est dit <u>régulier</u> si son centralisateur $\underline{\underline{Z}}(x)$ dans $\underline{\underline{G}}$ est de dimension minimale. Cette dimension est alors égale au rang de $\underline{\underline{G}}$ (noté rg $\underline{\underline{G}}$).

Lemme 12

Soit x un élément régulier de $\underline{\underline{\mathfrak{g}}}$ contenu dans $\mathfrak{b}$. Alors on a

$$\underline{\underline{Z}}(x) \subset \underline{\underline{B}}$$

Démonstration

Soit $\underline{\underline{Z}}_B(x)$ le centralisateur de x dans $\underline{\underline{B}}$. L'orbite de x par $\underline{\underline{B}}$ est contenue dans $x + \underline{\underline{\mathfrak{u}}}$, donc la dimension de $\underline{\underline{Z}}_B(x)$ est au plus égale à la dimension de $\mathfrak{b}/\mathfrak{u}$, c'est-à-dire au rang de $\underline{\underline{G}}$. Comme x est régulier la dimension de $\underline{\underline{Z}}_B(x)$ est au plus égale à rg $\underline{\underline{G}}$, donc finalement

$$\dim \underline{\underline{Z}}_B(x) = \mathrm{rg}\ \underline{\underline{G}} = \dim \underline{\underline{Z}}(x)$$

Comme, de plus, $\underline{\underline{Z}}(x)$ est connexe ([14], Prop. 14 ; comme $\underline{\underline{Z}}(x)$ et $\underline{\underline{B}}$ contiennent le centre de $\underline{\underline{G}}$, on peut, en considérant le quotient de ces groupes par le centre de $\underline{\underline{G}}$, se ramener au cas où $\underline{\underline{G}}$ est le groupe adjoint de $\underline{\underline{\mathfrak{g}}}$), on en déduit que $\underline{\underline{Z}}_B(x) = \underline{\underline{Z}}(x)$.

Lemme 13

Il existe un entier m_o tel que

$$(\mathrm{Ad}\ B_o)\,\bar{\mathfrak{b}} \supset e_- + \mathfrak{g}_{m_o} .$$

Démonstration

Il suffit de voir que $(\mathrm{Ad}\ B_o)\bar{\mathfrak{b}}$ est un voisinage de e_- dans $\mathfrak{g}$. Pour cela, montrons que l'application de $B_o \times \bar{\mathfrak{b}}$ dans $\mathfrak{g}$ qui envoie (g,x) sur (Ad g) x est une submersion au point (e,e_-), où e est l'élément neutre de G. L'application tangente en ce point est l'application de $\mathfrak{b} \times \bar{\mathfrak{b}}$ dans $\mathfrak{g}$ qui envoie (y,x) sur $[y,e_-] + x$.

Cette application a pour image $[b,e_-] + \overline{b}$, qui est égal à $\mathfrak{g}$ (cf. [14]). Elle est donc surjective.

Lemme 14

Soit m un entier au moins égal à m_o, soit x un élément de $e_- + \eta^{-1}(\mathfrak{a}_m)$ et Z(x) son centralisateur dans G, et soit u un élément de $U \cap G_m Z(x) G_m$. Alors u appartient à U_m.

Démonstration

D'après le lemme 13, il existe un élément g de B_o tel que $x \in (\mathrm{Ad}\ g)\overline{\mathfrak{b}}$. Posons $\mathfrak{b}' = (\mathrm{Ad}\ g)\overline{\mathfrak{b}}$ et soit $B' = g\,\overline{B}\,g^{-1}$. D'après [14], (théorème 0.10) x est régulier. D'après le lemme 12, on a donc $Z(x) \subset B'$. Par conséquent u appartient à $G_m B' G_m$. Or on a :

$$\begin{aligned} G_m &= g\, G_m\, g^{-1} \\ &= g\, U_m\, \overline{B}_m\, g^{-1} \\ &= U_m\, B'_m \end{aligned}$$

On a de même $G_m = B'_m U_m$. Donc on a :

$$G_m B' G_m = U_m B' U_m$$

On peut donc poser $u = u_1\, y\, u_2$ avec $u_i \in U_m$ et $y \in B'$. On a $u_1^{-1}\, u\, u_2^{-1} \in B' \cap U = \{e\}$. Donc $u = u_1 u_2$ et donc u est dans U_m.

V.4. Démonstration du lemme 4

Il suffit de montrer que, si u est un élément de U tel que $\psi_n * \varepsilon(u) * \psi_n \neq 0$, alors on a $u \in U_n$. En effet, si cette relation est vérifiée, alors les fonctions ψ_n et $\varepsilon(u) * \psi_n * \varepsilon(u^{-1})$ coïncident sur $G_n \cap u\, G_n\, u^{-1}$, et par conséquent les fonctions $\tilde{\psi}_n$ et $\tilde{\psi}_n \circ \mathrm{Ad}\, u^{-1}$ coïncident sur $\mathfrak{g}_n \cap (\mathrm{Ad}\ u)\mathfrak{g}_n$. Il existe donc un caractère χ de $\mathfrak{g}$ dont la restriction à $\mathfrak{g}_n$ (resp. à $(\mathrm{Ad}\ u)\mathfrak{g}_n$) est le caractère $\tilde{\psi}_n$ (resp. $\tilde{\psi}_n \circ \mathrm{Ad}\, u^{-1}$) de $\mathfrak{g}_n$ (resp. $(\mathrm{Ad}\ u)\mathfrak{g}_n$). D'après le lemme 9, on a donc

$$\chi = \mathrm{Ad}^* g_1\, (\varpi^{-2n}\tilde{\psi} + x_1)$$

$$(\mathrm{Ad}^* u^{-1})\chi = \mathrm{Ad}^* g_2\, (\varpi^{-2n}\tilde{\psi} + x_2)$$

où $g_i \in G_n$ et $x_i \in \mathfrak{a}_{-n}$. D'où

$$\varpi^{-2n}\tilde{\psi} + x_1 = \mathrm{Ad}^* (g_1^{-1} u\, g_2)(\varpi^{-2n}\tilde{\psi} + x_2)$$

D'après le lemme 11 ii), on a $x_1 = x_2$ donc $g_1^{-1}\, u\, g_2 \in Z\,(\varpi^{-2n}\,\tilde{\Psi} + x_1)$. On a donc $u \in U_n$ d'après le lemme 14, si n est assez grand.

VI - Cas du groupe GL_n.

Dans ce cas l'algèbre de Lie $\mathfrak{g}$ est égale à $\mathfrak{gl}_n(K)$. Dans la démonstration du lemme 4, on peut alors remplacer l'application exponentielle par l'application $x \longmapsto 1 + x$ de $\mathfrak{g}_1$ dans $GL_n(K)$. Les résultats du lemme 4, et des énoncés qui s'en déduisent sont alors vrais sans restriction sur K.

Références

[1] N. BOURBAKI, Algèbre, Chapitres 1 à 3, Hermann, Paris, 1970.

[2] N. BOURBAKI, Algèbre commutative, chapitres 3 et 4, Actu. Sci. Ind. n°1293, Hermann, Paris, 1962.

[3] B. BOURBAKI, Groupes et algèbres de Lie, Chapitres 2 et 3, Actu. Sci. Ind. n°1349, Hermann, Paris, 1972.

[4] F. BRUHAT, Sur une classe de sous-groupes compacts maximaux des groupes de Chevalley sur un corps p-adique, Publ. Math. I.H.E.S. 23 (1964), 45-74.

[5] F. BRUHAT et J. TITS, Groupes réductifs sur un corps local. I. Données radicielles valuées. Publ. Math. I.H.E.S. 41 (1972), 5-251.

[6] W. CASSELMAN, The Steinberg character as a true character, Proc. Sympos. Pure Math., vol. 26, Providence, R.I., 1974, 413-417.

[7] W. CASSELMAN, Some general results in the theory of admissible representations of p-adic reductive groups, à paraître.

[8] C. CHEVALLEY, Sur certains groupes simples, Tohoku Math. J. (2) 7 (1955), 14-66.

[9] P. GERARDIN, Sur les séries discrètes non ramifiées des groupes réductifs déployés p-adiques, thèse, Université Paris 7, 1974.

[10] HARISH-CHANDRA, Harmonic analysis on reductive p-adic groups, Proc. Sympos. Pure Math., vol. 26, Providence, R.I., 1974, 167-192.

[11] R.HOWE, The Fourier transform and germs of characters, notes polycopiées.

[12] R. HOWE, Kirillov theory for compact p-adic groups, notes polycopiées.

[13] H. JACQUET et R.P. LANGLANDS, Automorphic forms on GL(2), Lecture notes in Math. n°114, Springer-Verlag, Berlin et New-York, 1970.

[14] B. KOSTANT, Lie group representations on polynomial rings, Amer. J. Math. 85 (1963), 327-404.

[15] F. RODIER, Whittaker models for admissible representations of reductive p-adic split groups, Proc. Sympos. Pure Math., vol. 26, Providence R.I., 1974, 425-430.

[16] F. RODIER, Modèles de Whittaker des représentations admissibles des groupes réductifs p-adiques déployés, notes polycopiées (1972).

[17] J.A. SHALIKA, The multiplicity one theorem for GL(n), Annals of Math, vol. 100 (1974), 171-193.

[18] R. STEINBERG, Lectures on Chevalley groups, Yale University (1967).

[19] A. WEIL, Foundations of algebraic geometry, Amer. Math. Soc. Colloquium Publ. n°29, 2e éd., 1962.

Université Paris VII
U. E. R. de Mathématiques
2, place Jussieu

75221 PARIS CEDEX 05

SOME REMARKS ABOUT THE DISCRETE SERIES CHARACTERS OF $Sp(n,\mathbb{R})$

Wilfried Schmid*

Let G be a connected, semisimple Lie group, which admits a faithful finite-dimensional representation, and let K be a maximal compact subgroup. According to Harish-Chandra's criterion [1], G has a non-empty discrete series precisely when $\mathrm{rk}\, G = \mathrm{rk}\, K$. In this situation, one can choose a compact Cartan subgroup H of G, with $H \subset K$. I shall denote the Lie algebras of G, K, H by $\mathfrak{g}, \mathfrak{k}, \mathfrak{h}$, and their complexifications by $\mathfrak{g}^{\mathbb{C}}, \mathfrak{k}^{\mathbb{C}}, \mathfrak{h}^{\mathbb{C}}$. The dual group $\hat{H}$ of H is isomorphic, via exponentiation, to a lattice $\Lambda \subset i\mathfrak{h}^*$; for $\lambda \in \Lambda$, the corresponding character of H will be written as e^{λ}. The set of non-zero roots of $(\mathfrak{g}^{\mathbb{C}}, \mathfrak{h}^{\mathbb{C}})$, to which I shall refer as Φ, is contained in Λ.

For simplicity, I assume that the complexification $G^{\mathbb{C}}$ of G is simply connected. Harish-Chandra's enumeration of the discrete series [1] can then be rephrased as follows: for every nonsingular $\lambda \in \Lambda$ (i.e. $(\lambda,\alpha) \neq 0$ for all $\alpha \in \Phi$), there exists a unique tempered, invariant eigendistribution Θ_{λ}, such that

*Supported in part by the Sonderforschungsbereich 40 at the University of Bonn, and by the National Science Foundation, Grant GP 32843.

$$(1) \qquad \Theta_\lambda\Big|_H = (-1)^q \frac{\Sigma_{w\in W}\ \varepsilon(w)e^{w\lambda}}{\Pi_{\alpha\in\Phi,\,(\alpha,\lambda)>0}(e^{\alpha/2} - e^{-\alpha/2})} \quad ;$$

here W is the Weyl group of H in K, $\varepsilon(w)$ is the sign of w, and $q = \frac{1}{2}\dim G/K$. Two of these invariant eigendistributions, Θ_{λ_1} and Θ_{λ_2}, coincide precisely when λ_1 and λ_2 lie in the same W-orbit. Every Θ_λ is the character of a discrete series representation, and conversely.

The formula (1) only describes the discrete series characters on the elliptic set. It is therefore natural to ask whether the restrictions of the Θ_λ to the various other Cartan subgroups can also be expressed by a comparably simple formula. For groups of real rank one [2], for the indefinite unitary groups [4], and for $Sp(2,\mathbb{R})$ [5], such explicit formulas are known, and they do have a fairly simple appearance. This note is to suggest that the general situation can be a great deal more complicated. As evidence, I shall offer the real symplectic groups $Sp(n,\mathbb{R})$, whose discrete series characters I shall investigate by means of the algorithm given in [7].

To put the computations into perspective, some preliminary remarks are in order. For an arbitrary Cartan subgroup $B \subset G$, and for any given connected component B^j of B, there exists a certain semisimple subgroup S of G, with the following properties:

a) S contains both a split and a compact Cartan subgroup;

(2) b) the discrete series characters of G, restricted to B^j, can be expressed, in a simple and concrete manner, in terms of discrete series characters of S, restricted to the identity component of a split Cartan subgroup of S.

As was pointed out to me by G. Zuckerman, the relationship mentioned in b) follows from Harish-Chandra's construction of the discrete series characters. For groups G, whose quotient G/K carries a Hermitian symmetric structure, this relationship is stated as theorem (4.21) and theorem (4.22) in [7].

When G has real rank one, or when $G = SU(p,q)$, the subgroup S is necessarily a product of copies of $SL(2,\mathbb{R})$ and $SL(2,\mathbb{R})/\{\pm 1\}$. It is therefore not surprising that the discrete series characters of these groups have a relatively simple form. In view of (2), if one wants to compute the discrete series characters of an arbitrary group G, restricted to an arbitrary Cartan subgroup, it suffices to understand one rather special situation: the case of a split Cartan subgroup in group G which splits over $\mathbb{R}$. Apart from four exceptional cases, the only simple, algebraically simply connected matrix groups containing both a split and a compact Cartan subgroup are $Spin(2n,2n)$, $Spin(n,n+1)$, and $Sp(n,\mathbb{R})$. Of these, only the latter falls into the class of groups discussed in [7], i.e. the class of groups G with a Hermitian symmetric quotient G/K. However, as was remarked in §9 of [7], it seems likely that the arguments[1)] of [7] will eventually work for all semisimple matrix groups. The computations, which I shall carry out below for $Sp(n,\mathbb{R})$, would then have analogues for $Sp(2n,2n)$ and $Spin(n,n+1)$.

Before restricting my attention to $Sp(n,\mathbb{R})$, I shall recall a few facts from §4 and §9 of [7]. For this purpose, no assumptions about G are needed, beyond those made at the beginning of this paper. I shall write Φ^c and Φ^n for the sets of, respectively, compact and noncompact roots in Φ. Now let Ψ be a system of positive roots in Φ, and let

$$\Lambda_\Psi = \{\lambda \in \Lambda \mid (\lambda,\alpha) > 0 \quad \text{whenever} \quad \alpha \in \Psi\} \quad .$$

For any given Cartan subgroup $B \subset G$, on any particular connected component of the intersection of B with the regular set in G, the discrete series characters Θ_λ can be expressed as the quotient of two integral linear combinations of elements of $Hom(B,\mathbb{C}^*)$. The denominator, which coincides with the denominator of Weyl's character formula, does not depend on λ. As long as λ is restricted to lie in Λ_Ψ, in the formula for the numerator, λ only appears as a parameter; the formula continues to make sense for any $\lambda \in \Lambda$, even if $\lambda \notin \Lambda_\Psi$. These facts are implicit in Harish-Chandra's construction of the discrete series characters.

In view of the preceeding remarks, if one considers the formula for Θ_λ with $\lambda \in \Lambda_\Psi$, but letting λ wander over the

---

1) except, perhaps, for the proof of Blattner's conjecture.

larger set[2)]

(3) $$\{\lambda \in \Psi \mid (\lambda,\alpha) > 0 \quad \text{if} \quad \alpha \in \Psi \cap \Phi^c\} \quad ,$$

one obtains a collection of invariant eigendistributions $\Theta(\Psi,\lambda)$, depending on the choice of the system of positive roots Ψ, and parameterized by the set (3). As a consequence of the definition of the $\Theta(\Psi,\lambda)$,

(4) $$\Theta(\Psi,\lambda) = \Theta_\lambda \quad , \quad \text{if} \quad \lambda \in \Lambda_\Psi \quad .$$

Moreover,

(5) $$\Theta(w\Psi,w\lambda) = \Theta(\Psi,\lambda) \quad ,$$

for every $w \in W$ = Weyl group of H in K.

In order to describe a crucial relationship between the various $\Theta(\Psi,\lambda)$, I look at a system of positive roots Ψ, and a noncompact root $\beta \in \Phi$, which is simple, with respect to Ψ. The reflection about β determines an element s_β of the Weyl group of $(\mathfrak{g}^{\mathbb{C}},\mathfrak{h}^{\mathbb{C}})$. As is shown in §9 of [7], for any $\lambda \in \Lambda$ which has positive inner product with all roots α in $\Phi^c \cap \Psi = \Phi^c \cap s_\beta\Psi$,

(6) $$\Theta(\Psi,\lambda) + \Theta(s_\beta\Psi,\lambda) = \Theta \quad .$$

Here Θ stands for a certain induced invariant eigendistribution. I shall not repeat the precise definition of Θ, which is given

---

2) One could let λ wander over all of Λ, without obtaining more invariant eigendistributions, however. The parametrization by the set (3) has certain technical advantages.

in (4.15c) of [7]. It should be remarked, however, that Θ is induced from a maximal cuspidal parabolic subgroup. Furthermore, on the Levi component M of the parabolic subgroup, the inducing character belongs to the class of invariant eigendistributions $\Theta(\ldots,\ldots)$ of M.

Next, I assume that G/K can be given a Hermitian symmetric structure. In this situation, there exists a system of positive roots Ψ in Φ, such that

$$(7) \qquad \alpha_1, \alpha_2 \in \Psi \cap \Phi^n \text{ implies } \alpha_1 + \alpha_2 \notin \Phi .$$

For any system of positive roots Ψ with this property, the invariant eigendistributions $\Theta(\Psi,\lambda)$ were explicitely and globally computed by S. Martens [6] and H. Hecht [3]. A statement of their results can also be found in §3 of [7]. In effect, the global formula for $\Theta(\Psi,\lambda)$, with Ψ satisfying (7), coupled with the relationship (6), gives an algorithm for the computation of the discrete series characters, provided G/K has a Hermitian symmetric structure.

With these preliminaries out of the way, I limit my attention to the group $G = Sp(n,\mathbb{R})$. All other symbols retain their previous meaning. In $i\mathfrak{h}^*$, one can then pick an orthogonal basis $\{\beta_1,\ldots,\beta_n\}$, all of whose members have the same length, such that

$$(8) \qquad \begin{aligned} \Phi^n &= \{\pm\beta_i \,,\ 1 \leq i \leq n \,;\ \pm\gamma_{ij} \,,\ 1 \leq i < j \leq n\} \\ \Phi^c &= \{\pm\alpha_{ij} \,,\ 1 \leq i < j \leq n\} \,, \text{ with} \\ \alpha_{ij} &= \tfrac{1}{2}(\beta_i - \beta_j) \,,\quad \gamma_{ij} = \tfrac{1}{2}(\beta_i + \beta_j) \quad (1 \leq i < j \leq n) . \end{aligned}$$

As can be checked easily,

(9) Λ is the lattice spanned by $\frac{1}{2}\beta_1,\ldots,\frac{1}{2}\beta_n$.

Once and for all, I shall keep fixed the particular system of positive roots

(10) $$\Psi_0 = \{\alpha_{ij} , \gamma_{ij} , 1 \leq i < j \leq n ; \beta_i , 1 \leq i \leq n\} ;$$

Ψ_0 has the property (7).

I shall have to consider various subsets of $\Psi_0 \cap \Phi^n$, consisting of pairwise orthogonal roots. For the sake of brevity, such a set will be referred to as an orthogonal subset of $\Psi_0 \cap \Phi^n$. For any two noncompact roots γ,γ' , if $\gamma \perp \gamma'$, then γ and γ' are strongly orthogonal, i.e. $\gamma \pm \gamma' \notin \Phi$. In particular, every orthogonal subset of $\Phi^n \cap \Psi_0$ is composed of pairwise strongly orthogonal roots.

Let $S \subset \Phi^n \cap \Psi_0$ be an orthogonal subset. In the usual manner (cf. (2.12) of [7], for example), I associate to S a Cayley transform c_S ; it is an element of the adjoint group of $\mathfrak{g}^{\mathbb{C}}$. In $\mathfrak{g}$, there exists a Cartan subalgebra $\mathfrak{b}_S$, such that

(11) $$\mathfrak{b}_S^{\mathbb{C}} = c_S(\mathfrak{h}^{\mathbb{C}}) \quad .$$

The corresponding Cartan subgroup B_S of G has a decomposition

(12) $$B_S = B_{S,+} \cdot B_{S,-} \quad , \quad B_{S,+} = B_{S,+}^0 \cdot F_S \quad ;$$

here $B_{S,+}$ is a compact subgroup of B_S, with identity component $B_{S,+}^0$, $B_{S,-}$ is the split part of B_S, and F_S is a finite group, whose elements have order two. I denote the Lie algebras of $B_{S,+}$ and $B_{S,-}$ by $\mathfrak{b}_{S,+}$ and $\mathfrak{b}_{S,-}$, and I define

$$\begin{aligned} &\mathfrak{h}_S = \{X \in \mathfrak{h} \mid \langle \gamma, X\rangle = 0 \quad \text{if} \quad \gamma \in S\} \\ (13) \qquad &\mathfrak{h}_S^{\perp} = \text{orthogonal complement of } \mathfrak{h}_S \text{ in } \mathfrak{h}, \text{ relative} \\ &\qquad \text{to the Killing form of } \mathfrak{g}. \end{aligned}$$

As follows from the explicit definition of c_S,

$$\begin{aligned} &\mathfrak{b}_{S,+} = \mathfrak{h}_S \quad , \quad \mathfrak{b}_{S,-} = c_S(i\mathfrak{h}_S^{\perp}) \quad , \\ (14) \qquad &F_S = \{h \in \exp(\mathfrak{h}_S^{\perp}) \mid e^{\lambda}(h) = \pm 1 \quad \text{whenever} \quad \lambda \in \Lambda\} \quad . \end{aligned}$$

Further notation: M_S shall stand for the centralizer of $B_{S,-}$ in G, M_S^o for the identity component of M_S, and M_S^+ for the intermediate subgroup

$$M_S^+ = \{m \in M_S \mid \text{Ad } m\colon M_S^o \to M_S^o \quad \text{is inner}\} \quad .$$

Then F_S is central in M_S^+, and $M_S^+ = M_S^o \cdot F_S$.

The Lie algebra $\mathfrak{m}_S$ of M_S^o contains $\mathfrak{b}_{S,+} = \mathfrak{h}_S$ as a Cartan subalgebra, and hence H intersects M_S in a compact Cartan subgroup of M_S. In a natural fashion, the root system of $(\mathfrak{m}_S^{\mathbb{C}}, \mathfrak{b}_{S,+}^{\mathbb{C}})$ can be identified with

$$(15) \qquad \Phi_S = \{\gamma \in \Phi \mid \gamma \perp S\} \quad .$$

By Φ_S^c and Φ_S^n, I denote the sets of those roots in Φ_S, which are, respectively, compact and noncompact, viewed as roots of $(M_S^o, M_S^o \cap H)$. It should be realized that Φ_S^n need not coincide with $\Phi_S \cap \Phi^n$. Nevertheless, the system of positive roots

$\Psi_o \cap \Phi_S$ in Φ_S inherits the property (7) from Ψ_o:

(16) $\gamma, \gamma' \in \Psi_o \cap \Phi_S^n$ implies $\gamma + \gamma' \notin \Phi_S$.

Also, if $\lambda \in \Lambda$ has the property

(17) $(\lambda, \alpha) > 0$ if $\alpha \in \Phi^c \cap \Psi_o$,

then λ satisfies the condition

(18) $(\lambda, \alpha) > 0$ whenever $\alpha \in \Phi_S^c \cap \Psi_o$.

In order to verify these statements, one should observe that

(19) $M_S^o \simeq Sp(n-k-2\ell, \mathbb{R}) \times SL(2,\mathbb{R}) \times \ldots \times SL(2,\mathbb{R})$

(ℓ copies of $SL(2,\mathbb{R})$), with k = number of long roots in S, ℓ = number of short roots in S.

In addition to the orthogonal subset S of $\Phi^n \cap \Psi_o$, I now consider a particular $\lambda \in \Lambda$, subject to the condition (17). Corresponding to each such pair S, λ , I shall introduce an invariant eigendistribution $\Theta_S(\lambda)$, as will be described next. Because of (18), the system of positive roots $\Psi_o \cap \Phi_S$ in Φ_S, together with the restriction of λ to $\mathfrak{h}_S$, determine an invariant eigendistribution

$$\varphi_o = \Theta(\Psi_o \cap \Phi_S , \lambda \big|_{\mathfrak{h}_S})$$

on M_S^o. Since $\Psi_o \cap \Phi_S$ satisfies (16), φ_o can be explicitly and globally computed. The elements of the set S can be enumerated as

(20)
$$\beta_{k_1},\beta_{k_2},\ldots,\beta_{k_s}\ ;\ \gamma_{\ell_1 m_1},\gamma_{\ell_2 m_2},\ldots,\gamma_{\ell_t m_t}\ , \quad \text{with}$$
$$1 \le k_s < \ldots < k_1 \le n\ ,\ 1 \le \ell_t < \ldots < \ell_1 \le n\ ,\ \ell_j < m_j\ .$$

Let $\xi_S\colon F_S \to \{\pm 1\}$ be the homomorphism given by

$$\xi_S(f) = e^{\mu}(f)\ , \quad \text{for } f \in F_S$$

(recall: $F_S \subset H^!$), where

$$\mu = \lambda + \frac{1}{2}\Sigma_{j=1}^{s}\ (n+1-j)\beta_{k_j} + \frac{1}{2}\Sigma_{j=1}^{t}\ \beta_{\ell_j}\ .$$

One can extend φ_o to an invariant eigendistribution φ_1 on M_S^+, and then further to an invariant eigendistribution φ_2 on M_S, by setting

$$\varphi_1(mf) = \varphi_o(m)\xi_S(f) \qquad (m \in M_S^o\ ,\ f \in F_S)\ ,$$

$$\varphi_2 = \begin{cases} \Sigma_{m\in M_S/M_S^+}\ \varphi_1\circ \mathrm{Ad}\ m\ , & \text{on } M_S^+ \\ 0\ , & \text{on the complement of } M_S^+ \text{ in } M_S\ . \end{cases}$$

Via the Cayley transform c_S, λ determines a linear functional $\tilde{\lambda}$ on b_S^C, whose restriction to $b_{S,-}$ shall be denoted by ν. Putting together φ_2 and e^{ν}, one obtains an invariant eigendistribution φ on $M_S B_{S,-}$:

$$\varphi(mb) = \varphi_2(m)e^{\nu}(b) \qquad (m \in M_S,\ b \in B_{S,-})\ .$$

Now let $P_S = M_S B_{S,-} N$ be a cuspidal parabolic subgroup associated to the Cartan subgroup B_S. In the customary manner, I extend φ

to P_S, and I induce the extension from P_S to G. The result of this construction is an invariant eigendistribution $\Theta_S(\lambda)$.

One more ingredient is needed for the statement of theorem 1 below. For any two distinct integers ℓ, m between 1 and n, I set

$$\varepsilon(\ell,m) = \begin{cases} +1 , & \text{if } \ell < m \\ -1 , & \text{if } \ell > m . \end{cases}$$

If $S \subset \Psi_o \cap \Phi^n$ is an orthogonal subset, whose elements have been enumerated as in (19), I define

$$(21) \quad \begin{aligned} \varepsilon(S) = (-1)^{s(n+1)+t} \, & \Pi_{i=1}^{s}\{(-1)^{k_i} \Pi_{j=1}^{t}(\varepsilon(k_i,\ell_j)\varepsilon(k_i,m_j))\} \\ \times \, & \Pi_{i=1}^{t}\{(-1)^{\ell_i+m_i} \Pi_{j=i+1}^{t}(\varepsilon(\ell_i,\ell_j)\varepsilon(\ell_i,m_j)\varepsilon(m_i,\ell_j)\varepsilon(m_i,m_j))\}. \end{aligned}$$

Thus $\varepsilon(S) = \pm 1$, and $\varepsilon(\emptyset) = +1$.

For any system of positive roots $\Psi \subset \Phi$, there exists a unique element $w \in W$, which transforms $\Psi \cap \Phi^c$ into $\Psi_o \cap \Phi^c$. In other words, every such system of positive roots Ψ is W-conjugate to one with the property

$$(22) \qquad \Psi \cap \Phi^c = \Psi_o \cap \Phi^c .$$

Hence, and in view of (5), if one wants to have global formulas for the invariant eigendistributions $\Theta(\Psi,\lambda)$ -- and thus for the discrete series characters -- , it suffices to consider systems of positive roots satisfying (22). Clearly (22) implies the containment

$$\Psi_o \cap (-\Psi) \subset \Psi_o \cap \Phi^n \quad .$$

Now let $\Psi \subset \Phi$ be an arbitrary system of positive roots with the property (22), and suppose that $\lambda \in \Lambda$ belongs to the set (3). In this situation, $\Theta(\Psi,\lambda)$ is well-defined. Also, λ then satisfies the condition (17). I set N = cardinality of $\Psi_o \cap (-\Psi)$.

Theorem 1. Under the hypothesis stated above,

$$\Theta(\Psi,\lambda) = (-1)^N \Sigma_S \varepsilon(S) \Theta_S(\lambda) \quad ,$$

with S running over all orthogonal subsets of $\Psi_o \cap (-\Psi)$, including the empty set.

If $\Psi = \Psi_o$, the identity above becomes a tautology. Also, if Ψ satisfies (22) and is otherwise arbitrary, there exists a chain of systems of positive roots $\Psi_o, \Psi_1, \ldots, \Psi_m$, with $\Psi_m = \Psi$, such that each Ψ_i is obtained from Ψ_{i-1} by reflection about a noncompact, simple (relative to Ψ_{i-1}) root. All of the Ψ_i then have the property (22). One can therefore prove the theorem by induction on m, simply by verifying that the theorem is consistent with the relationship (6). In view of (19), the induction step can proceed in turn by induction on the rank of $SP(n,R)$. Roughly speaking, the argument comes down to the fact that the process of inducing an invariant eigendistribution from a parabolic subgroup is transitive, in the appropriate sense. The details of the proof will be left to the interested reader.

Theorem 1 expresses each $\Theta(\Psi,\lambda)$ in terms of invariant eigendistributions which are completely and globally known, since the process of inducing an invariant eigendistribution can be carried out quite explicitly. Hence, in principle, the theorem leads to concrete formulas for the $\Theta(\Psi,\lambda)$, restricted to any given Cartan subgroup. Unfortunately, the resulting formulas are less than fully transparent.

In the context of $G = Sp(n,\mathbb{R})$, the subgroups mentioned in (2) turn out to be products of copies of $Sp(k,\mathbb{R})$, for various indices $k \leq n$. Hence, if one wants to compute the discrete series characters -- or equivalently, the invariant eigendistribution $\Theta(\Psi,\lambda)$ --, restricted to an arbitrary Cartan subgroup of G, it suffices to understand the discrete series characters of $Sp(n,\mathbb{R})$, restricted to the identity component of a split Cartan subgroup of G, with Lie algebra $\mathfrak{a}$, and with identity component $A^o = \exp \mathfrak{a}$. I choose an inner automorphism c of $\mathfrak{g}^{\mathbb{C}}$, such that

(23) $$c: \mathfrak{h}^{\mathbb{C}} \to \mathfrak{a}^{\mathbb{C}} \quad .$$

The Weyl group of A in G operates transitively on the set of Weyl chambers in $\mathfrak{a}$. Hence the restriction of any invariant eigendistribution to A^o is known, as soon as one knows it on

(24) $$C = \{a \in A^o \mid a = \exp(cX),\ X \in i\mathfrak{h},\ \langle\alpha,X\rangle < 0 \text{ if } \alpha \in \Psi_o\} \; .$$

I shall now calculate the invariant eigendistributions $\Theta_S(\lambda)$, or rather, their restrictions to A^o. For this purpose, I keep fixed a particular orthogonal subset S of $\Psi_o \cap \Phi^n$.

Without loss of generality, I assume that

$$A \subset M_S B_{S,-} \quad .$$

Let W_A be the Weyl group of A in G, and W'_A the Weyl group of A in $M_S B_{S,-}$. By Φ_A, I shall denote the root system of $(\mathfrak{g},\mathfrak{a})$. The subgroup $M_S B_{S,-}$ determines a sub-root system $\Phi'_A \subset \Phi_A$. Via c, Ψ_o cuts out a system of positive roots Ψ_A in Φ_A. As follows from (24),

(25) $$\alpha \in \Phi_A \text{ belongs to } \Psi_A \Leftrightarrow e^{\alpha} < 1 \text{ on } C \quad .$$

Appealing to theorem 4.3.8 of [8], for example, one may conclude[3)] that

(26) $$\Theta_S(\lambda) \mid_{A^o} = (\# W'_A)^{-1} \Sigma_{w \in W_A} \mid \Pi_{\alpha \in \Psi_A, \alpha \notin \Phi'_A} (e^{w\alpha/2} - e^{-w\alpha/2}) \mid^{-1} (\varphi \Big|_{A^o}) \circ w^{-1};$$

here φ has the same meaning as in the definition of $\Theta_S(\lambda)$. In analogy to (24), I define

(27) $$C' = \{a \in A^o \mid a = \exp(cX),\ X \in i\mathfrak{h},\ \langle \alpha, X \rangle < 0 \text{ if } \alpha \in \Phi_S \cap \Psi_o\}$$

(cf. (15)). I enumerate the elements of the set

(28) $$\{w \in W_A \mid w^{-1}C \subset C'\}$$

---

3) theorem 4.3.8 of [8] fails to take into account the possibility that several non-conjugate Cartan subgroups of $M_S B_{S,-}$ may become conjugate under G; however, since A splits, this is not a problem in the present context.

as $\{w_1,\ldots,w_r\}$. It is a set of representatives for the quotient W_A/W_A'. Hence, in (26), instead of summing over W_A, one may sum over this set, provided the factor $(\#W_A')^{-1}$ is dropped. For $\alpha \in \Phi_A$, let $\operatorname{sgn} \alpha = +1$ or -1, depending on whether $\alpha \in \Psi_A$ or $\alpha \notin \Psi_A$. Thus (26) becomes

$$\Theta_S(\lambda)\big|_C = \{\Sigma_{i=1}^r (\Pi_{\alpha\in\Psi_A,\alpha\notin\Phi_A'} \operatorname{sgn}(-w_i\alpha)) \times \tag{29}$$
$$(\Pi_{\alpha\in\Psi_A,\alpha\notin\Phi_A'} (e^{w_i\alpha/2} - e^{-w_i\alpha/2}))^{-1} (\varphi\big|_{C'})\circ w_i^{-1}\}\big|_C$$

(recall (25)!).

Every coset in M_S/M_S^+ has a representative m which normalizes A. Such an m, via Ad, operates on A as an element of W_A'; hence

$$\varphi_1\big|_A = \#(M_S/M_S^+)\varphi_o\big|_A \quad . \tag{30}$$

The inner automorphism c of $\mathfrak{g}^{\mathbb{C}}$ lifts to an automorphism of $G^{\mathbb{C}}$, which will also be referred to as c. Its inverse maps C and C' onto subsets of $H^{\mathbb{C}}$. Instead of describing $\Theta_S(\lambda)\big|_C$ and $\varphi\big|_{C'}$ directly, it will be more convenient to compose these functions with c^{-1}. Let W' be the subgroup of the Weyl group of $(\mathfrak{g}^{\mathbb{C}},\mathfrak{h}^{\mathbb{C}})$ which is generated by the reflections about the roots in Φ_S^c (Φ_S^c was defined below (15)); it can be checked that $W' \subset W$. The explicit formula for φ_o (cf. §3 of [7], for example), coupled with (30), leads to the statement on the subset $c^{-1}(C')$ of $H^{\mathbb{C}}$, $(\varphi|_{C'})\circ c^{-1}$ is given by

(31)
$$\#(M_S/M_S^+)(-1)^{q'}(\Pi_{\alpha\in\Phi_S\cap\Psi_o}(e^{\alpha/2}-e^{-\alpha/2}))^{-1}\Sigma_{w\in W'}\varepsilon(w)e^{w\lambda} \quad ,$$

with $q' = \frac{1}{2}\dim_R M_S^o/M_S^o \cap K$. For every $\mu \in \Lambda$, e^{μ} can be interpreted as a function on $H^{\mathbb{C}}$; the formula appearing in (31) is to be understood in this sense.

If S is the set (20),

(32)
$$\#(M_S/M_S^+) = 2^t$$

(cf. lemma (2.58) of [7], for example). Also,

$$q' = \frac{1}{2}(n-s-2t)(n-s-2t+1)+t \quad , \quad \text{and}$$

$$\#\Psi_o - \#(\Psi_o \cap \Phi_S) = n^2-(n-s-2t)^2-t \quad ,$$

so that

(33)
$$q' + \#\Psi_o - \#(\Psi_o \cap \Phi_S) \equiv \frac{1}{2}n(n+1)+\frac{1}{s}s(s+1)+(n+1)(s+t)+st \mod 2.$$

Let $W_{\mathbb{C}}$ denote the Weyl group of $(\mathfrak{g}^{\mathbb{C}},\mathfrak{h}^{\mathbb{C}})$. Under c, the subset (28) of W_A corresponds to

(34)
$$\{w \in W_{\mathbb{C}} \mid w(\Phi_S \cap \Psi_o) \subset \Psi_o\} \quad .$$

As w' and w'' run over, respectively, the group W' and the set (34), the products $w''w'$ exhaust the set

(35)
$$R_S=\{w \in W_{\mathbb{C}} \mid w\beta_i \in \Psi_o \text{ if } \beta_i \perp S,\ w\alpha_{\ell_j m_j} \in \Psi_o,\ 1 \le j \le t\}.$$

Combining (29-33), one now finds that $\Theta_S(\lambda)\big|_C$, pulled back to a function on $c^{-1}(C) \subset H^{\mathbb{C}}$ via c^{-1}, equals

$$(36)\quad (-1)^{\frac{1}{2}n(n+1)+\frac{1}{2}s(s+1)+(n+1)(s+t)+st} \, (\Pi_{\alpha\in\Psi_o}(e^{\alpha/2} - e^{-\alpha/2}))^{-1} \times \Sigma_{w\in R_S} (\Pi_{\alpha\in\Psi_o,\alpha\notin\Phi_S} \operatorname{sgn}(w\alpha))\,\varepsilon(w)\,e^{w\lambda} \; .$$

Here $\operatorname{sgn}(\alpha)$, for $\alpha \in \Phi$, denotes the sign of the root α, relative to the system of positive roots Ψ_o.

I now make the same hypothesis as above the statement of theorem 1.

<u>Theorem 2</u>. The invariant eigendistribution $\Theta(\Psi,\lambda)$, restricted to the cone $C \subset A^o$ and pulled back via c^{-1} to the subset $c^{-1}(C)$ of $H^{\mathbb{C}}$, is given by the formula

$$(\Pi_{\alpha\in\Psi_o}(e^{\alpha/2} - e^{-\alpha/2}))^{-1} \, \Sigma_S\{(-1)^{N+\frac{1}{2}n(n+1)+\frac{1}{2}s(s+1)+(n+1)(s+t)+st} \times$$

$$\varepsilon(S)\, \Sigma_{w\in R_S} (\Pi_{\alpha\in\Psi_o,\alpha\notin\Phi_S} \operatorname{sgn}(w\alpha))\; \varepsilon(w)e^{w\lambda} \; ;$$

S runs over all orthogonal subsets of $\Psi_o \cap (-\Psi)$, including the empty set.

The formula contained in the theorem is not as explicit as one would like, of course. If $(\lambda,\alpha) > 0$ for all roots $\alpha \in \Psi$, $\Theta(\Psi,\lambda)$ must be tempered. In particular, this implies that many of the terms in the formula cancel away. Except for some special choices of Ψ, it seems exceedingly difficult to carry out the cancellation - if it is possible at all. However, the formula has some special features which should be mentioned. For this purpose, I fix the choice of Ψ, subject to the condition (22). Let $\alpha_1,\ldots,\alpha_\ell$ be an enumeration of the simple roots in Ψ which are compact. These roots span a sub-root system Φ' of Φ^c. Let U be the subgroup of K which contains H, and which corresponds to the root system Φ'. Then U coincides with K if and only if Ψ has the property (7). As can be shown, the formula in theorem 2 is equal to an integral linear combination of irreducible characters of U, divided by

$$\Pi_{\alpha\in\Psi_o,\alpha\notin\Phi'} (e^{\alpha/2} - e^{-\alpha/2}) \quad .$$

When $U = K$, or equivalently, when Ψ satisfies condition (7), $\Theta(\Psi,\lambda)\Big|_{A^o}$ is given by a known and relatively simple formula. It turns out that for other choices of Ψ, as the semisimple rank of U becomes smaller, the explicit formula for $\Theta(\Psi,\lambda)\Big|_{A^o}$ becomes more complicated. I shall now look at those systems of positive roots Ψ for which U has semisimple rank n-2 (note: K has semisimple rank n-1).

I choose and keep fixed an integer k between 0 and n, and I define

(37)
$$\rho_k = \frac{1}{2}\,\Sigma_{j=1}^{n-k}\,(n+1-j)\beta_j - \frac{1}{2}\,\Sigma_{j=1}^{k}\, j\,\beta_{n-k+j} \quad ,$$
$$\Psi_k = \{\alpha \in \Phi \mid (\alpha,\rho_k) > 0\} \quad .$$

For $k = 0$, this definition is consistent with the previous choice of Ψ_o. Also, for $0 \leq k \leq n$,

$$\Psi_k \cap \Phi^c = \Psi_o \cap \Phi^c \quad ;$$

thus all the Ψ_k have the property (22). There exists an outer automorphism τ of G, which leaves H invariant, and which induces the mapping

$$\beta_j \to -\beta_{n-j} \quad , \quad 1 \leq j \leq n \,, \tag{38}$$

on $\mathfrak{h}^*$. As can be checked, the Ψ_k and $\tau\Psi_k$, with $0 < k < n$, are exactly those systems of positive roots for which U has semisimple rank n - 2, and which satisfy (22). Similarly, $\Psi_o = \tau\Psi_k$ and $\Psi_k = \tau\Psi_o$ are the only systems of positive roots satisfying (22), and with U = K. For any Ψ,

$$\Theta(\tau\Psi, \tau\lambda) \circ \tau = \Theta(\Psi, \lambda) \quad ;$$

Hence if one wants to compute $\Theta(\tau\Psi_k, \lambda)$, it suffices to know $\Theta(\Psi_k, \lambda)$.

In order to state an explicit formula for $\Theta(\Psi_k, \lambda)$, I shall need some notation. As before, W stands for the Weyl group of H in K. Viewed as a transformation group acting on $\mathfrak{h}^*$, W consists precisely of the permutations of $\beta_1, \ldots, \beta_n$. For any two distinct integers ℓ_1, ℓ_2 between 1 and n, I set

$$\varepsilon(w; \ell_1, \ell_2) = \begin{cases} +1 & \text{if } \ w(\frac{1}{2}(\beta_{\ell_1} - \beta_{\ell_2})) \in \Psi_o \\ -1 & \text{if } -w(\frac{1}{2}(\beta_{\ell_1} - \beta_{\ell_2})) \in \Psi_o \end{cases} \quad , \tag{39}$$

and I define

$$(40) \qquad \delta(w;\ell) = (-1)^{\ell'} \quad \text{if} \quad w\beta_\ell = \beta_{\ell'} \qquad (1 \le \ell \le n) \quad .$$

Furthermore, if $1 \le \ell \le n$, $s_\ell \in W_{\mathbb{C}}$ shall denote the reflection about the root β_ℓ. Now let k be given, $0 \le k \le n$, and suppose that $\lambda \in \Lambda$ is chosen subject to the condition (17).

<u>Theorem 3</u>. The invariant eigendistribution $\Theta(\Psi_k,\lambda)$, restricted to $C \subset A^o$ and pulled back via c^{-1} to the subset $c^{-1}(C)$ of $H^{\mathbb{C}}$, is given by the formula

$$(-1)^{\frac{1}{2}n(n+1)+\frac{1}{2}k(k+1)} (\Pi_{\alpha\in\Psi_o}(e^{\alpha/2} - e^{-\alpha/2}))^{-1} \Sigma_{r=0}^{k} (-1)^{\frac{1}{2}r(r-1)} \Sigma_{w\in W}$$

$$\Sigma_{n-k<\ell_1<\ldots<\ell_r\le n} \{\Pi_{n-k<\ell\le n, \ell\ne\ell_i} (1 + \delta(w;\ell)\Pi_{i=1}^{r} \ \epsilon(w;\ell,\ell_i)) \times$$

$$\Pi_{i=1}^{r} \ \delta(w;\ell_i) \ \epsilon(w) \ \exp(w(\Pi_{i=1}^{r} s_{\ell_i})\lambda)\} \quad .$$

This statement can be deduced from theorem 2. However, it is simpler to prove it directly, by double induction on n and k, using the identity (6). Many of the computations which enter the argument are similar to those in the proof of theorem 2. The missing details can be supplied by the interested reader.

The formula supplied by theorem 3 is completely explicit, of course. Although it can be written down in closed form without overwhelming difficulty, for general k, it looks considerably more complicated than for k = 0 or k = n. The coefficient of any exponential term appearing in the formula equals, up to sign, a power of two. This feature is special to the systems of positive roots Ψ_k. I shall conclude with a final example, which will support the preceeding statement. The example will also show that the explicit formula for $\Theta(\Psi,\lambda)$ becomes more complicated still as the semisimple rank of U drops to n-3.

I fix an integer k, with $0 < k < n$. With respect to Ψ_k, $\gamma_{n-k,n}$ is a noncompact, simple root (cf. (8)). Hence

$$\Psi_k' = (\Psi_k - \{\gamma_{n-k,n}\}) \cup \{-\gamma_{n-k,n}\}$$

is a system of positive roots, satisfying (22). The group U, which corresponds to Ψ_k', has semisimple rank n-3, provided $k \geq 2$. For any given $\lambda \in \Lambda$, such that (17) holds, I set

$$\Theta = \Theta(\Psi_k,\lambda) + \Theta(\Psi_k',\lambda) \quad . \tag{41}$$

As explained below (6), Θ is an induced invariant eigendistribution. The precise description of Θ, coupled with theorem 2, applied to the group $Sp(n-2,\mathbb{R})$, leads to an explicit formula for $\Theta\big|_{A^o}$: the restriction of Θ to $C \subset A^o$, pulled back to $c^{-1}(C) \subset H^{\mathbb{C}}$ via c^{-1}, is given by

$$(-1)^{\frac{1}{2}n(n+1)+\frac{1}{2}k(k+1)+1} (\Pi_{\alpha\in\Psi_o}(e^{\alpha/2}-e^{-\alpha/2}))^{-1} \Sigma_{r=0}^{k-1}(-1)^{\frac{1}{2}r(r-1)}$$

$$\Sigma_{n-k<\ell_1<\cdots<\ell_r<n}\ \Sigma_{w\in W}\ \{\Pi_{n-k<\ell<n,\,\ell\neq\ell_i}(1+\delta(w;\ell)\ \times$$

$$\varepsilon(w;\ell,n-k)\,\varepsilon(w;\ell,n)\,\Pi_{i=1}^{r}\ \varepsilon(w;\ell,\ell_i))\,\delta(w;n-k)\,\delta(w;n)\ \times$$

(42)
$$\Pi_{i=1}^{r}(\delta(w;\ell_i)\,\varepsilon(w;\ell_i,n-k)\,\varepsilon(w;\ell_i,n))\,\varepsilon(w)\ \times$$

$$[(1+\varepsilon(w;n-k,n))\exp(w(\Pi_{i=1}^{r}\ s_{\ell_i})\lambda)\ -\ (1-\varepsilon(w;n-k,n))\ \times$$

$$\exp(ws_{n-k}s_n(\Pi_{i=1}^{r}\ s_{\ell_i})\lambda)\ +\ 2\varepsilon(w;n-k,n)\exp(ws_n(\Pi_{i=1}^{r}\ s_{\ell_i})\lambda)]\}.$$

If one combines the formula (42) with theorem 3, one obtains a formula for $\Theta(\Psi_k',\lambda)$. Even though there will be some cancelation, the resulting formula becomes quite complicated. As is not hard to check, the coefficients of the various exponential terms assume a wide range of values; in particular, depending on the choices of k and n, odd coefficients do occur.

References

[1] Harish-Chandra, Discrete series II, Acta Math., 116(1966), 1-111.

[2] Harish-Chandra, Two theorems on semisimple Lie groups, Ann. of Math., 83(1966), 74-128.

[3] Hecht, H., The characters of Harish-Chandra representations, thesis, Columbia University, 1974, To appear.

[4] Hirai, T., Invariant eigendistributions of Laplace operators on semisimple Lie groups I. Case of SU(p,q), Japan J. of Math. 40(1970), 1-68.

[5] Hirai, T., Explicit form of the characters of discrete series representations of semisimple Lie groups, Proceedings of symposia in pure mathematics, Vol. XXVI, 281-287.

[6] Martens, S., The characters of the holomorphic discrete series, thesis, Columbia University, 1973.

[7] Schmid, W., On the characters of the discrete series of a semisimple Lie group (the Hermitian symmetric case), to appear.

[8] Wolf, J.A., Unitary representations on partially holomorphic cohomology spaces, Amer. Math. Soc. Memoir, 138(1974).

AN APPLICATION OF POLARISATIONS AND HALF-FORMS

D. J. Simms

INTRODUCTION

The concept of a polarisation of a symplectic manifold was introduced by Kostant in order to provide a unified framework for the construction of irreducible unitary representations of Lie groups which would generalise Kirillov's theory for nilpotent groups and the Borel-Weil theory for compact groups. The further introduction of half-forms normal to a polarisation was made in joint work [1] with Blattner and Sternberg in order to construct an operator intertwining the Hilbert spaces based on two different polarisations.

More recently these methods have also been applied to the symplectic manifolds which represent the phase spaces of simple physical systems. It is the purpose of this paper to describe one of these applications, with an outline of the proof. The example is of some interest in that it shows that the spectrum of an operator obtained from a classical function by 'quantisation' can be independent of the choice of polarisation. The method described in this paper differs considerably from that of [3] in that the quantisation procedure is applied directly to the original symplectic manifold instead of to quotients of submanifolds.

THE KEPLER PROBLEM

Let $(\mathbb{R}^3 - \{0\}) \times \mathbb{R}^3$ be the cotangent bundle of $\mathbb{R}^3 - \{0\}$ with coordinates q_j, p_j, $j = 1, 2, 3$, and symplectic form $\sum_{j=1}^{3} dp_j \wedge dq_j$. Let $H = \frac{1}{2m} p^2 - \frac{K}{q}$ where $p^2 = \sum_{j=1}^{3} p_j^2$, $q^2 = \sum_{j=1}^{3} q_j^2$, $K > 0$, $m > 0$. Let $M = H^{-1}(-\infty, 0)$ and let ω be the restriction to M of the symplectic form. Let ξ_H be the Hamiltonian vector field given by

$$dH = \omega(\xi_H, \cdot) .$$

Let F be an involutory sub-bundle of the complexified tangent bundle of M, maximally isotropic with respect to ω. Let $\mathcal{U}_F$ denote the space of smooth sections of F and suppose F is chosen so that $\xi_H \in \mathcal{U}_F$.

Thus M is the negative energy phase space of a particle of mass m under an inverse square central force (the Kepler problem), H is the energy, ξ_H is the vector field generated by the classical motion, and F is a polarisation of the symplectic manifold (M, ω) which includes the direction of the classical motion.

Let (L, ∇) be the unique hermitian line bundle over M having ω as curvature form; see [2]. Let L^F be the complex line bundle of half-forms on M normal to F, and let ∇ also denote the unique covariant derivative of sections of L^F along vector fields in $\mathcal{U}_F$ which vanishes on half-forms dual to frames consisting of Hamiltonian vector fields. Then for each $\xi \in \mathcal{U}_F$ we have an operator ∇_ξ of covariant differentiation acting on the space $D'(L \otimes L^F)$ of distributional sections of $L \otimes L^F$. Let

$$\mathcal{W}^F = \{\psi \in D'(L \otimes L^F) \mid \nabla_\xi \psi = 0 \qquad \text{all} \quad \xi \in \mathcal{U}_F\}$$

be the space of distributional sections polarised along F.

The classical motion, generated by the vector field ξ_H, gives a 1-dimensional foliation of M (suitably completed) with closed leaves. Denote by M_o the set of points of M at which the holonomy group of (L, ∇) around the corresponding leaf is trivial. Let $\tilde{M}$ denote the quotient of M_o by the foliation. The symplectic form ω on M induces a symplectic form $\tilde{\omega}$ on $\tilde{M}$, and the polarisation F projects to a polarisation $\tilde{F}$ of $\tilde{\omega}$. We define an equivalence relation on $L|M_o$ by parallel transport along the leaves of the foliation. Let $\tilde{L}$ be the quotient line bundle over $\tilde{M}$ and let $L^{\tilde{F}}$ be the line bundle of half-forms on $\tilde{M}$ normal to $\tilde{F}$. The connection on L induces a connection on $\tilde{L}$ and hence we have an operator ∇_ξ of covariant differentiation acting on the space $D'(\tilde{L} \otimes L^{\tilde{F}})$ of distributional sections of $\tilde{L} \otimes L^{\tilde{F}}$ for all $\xi \in \mathcal{U}_{\tilde{F}}$.

The elements of $\mathcal{W}^F$ have support in M_o and may be identified with the space

$$\{\psi \in D'(\tilde{L} \otimes L^{\tilde{F}}) \mid \nabla_\xi \psi = 0 \qquad \text{all} \quad \xi \in \mathcal{U}_{\tilde{F}}\}$$

of distributional sections of $\tilde{L} \otimes L^{\tilde{F}}$ polarised along $\tilde{F}$. Now M_o is the union over positive integers n of the submanifolds $M^{(n)} = H^{-1}(-2\pi^2 m K^2 n^{-2})$. The quotient $\tilde{M}^{(n)}$ of $M^{(n)}$ is diffeomorphic to $S^2 \times S^2$ and the restriction $\tilde{\omega}^{(n)}$

of $\tilde{\omega}$ to $\tilde{M}^{(n)}$ has de Rham cohomology class $n(a_1 + a_2)$ where a_1, a_2 are the standard generators of $H^*(S^2 \times S^2, \mathbb{R})$. The polarisation $\tilde{F}$ necessarily defines a complex structure on $\tilde{M}$ and we make the additional assumption that the resulting Riemannian structure on $\tilde{M}$ is positive definite. Thus each $\tilde{M}^{(n)}$ is a compact Kähler manifold. It follows that W^F may be identified with the space of holomorphic sections of $\tilde{L} \otimes L^{\tilde{F}}$ with respect to a natural holomorphic structure on $\tilde{L} \otimes L^{\tilde{F}}$.

Thus

$$W^F = \oplus H^{(n)}$$

where $H^{(n)}$ is the space of holomorphic sections of $\tilde{L} \otimes L^{\tilde{F}}$ over $\tilde{M}^{(n)}$. The Chern class of the complex line bundle $L^{\tilde{F}}$ is half the Chern class of the canonical line bundle of the complex structure $\tilde{F}$. An application of the Riemann-Roch theorem shows that $H^{(n)}$ has complex dimension n^2.

The function $H \in C^\infty(M)$ defines an operator on the sections of $L \otimes L^F$ by Kostant's quantisation procedure. The operator induced on W^F has $H^{(n)}$ as eigenspace with eigenvalue $\frac{-2\pi^2 m\ K^2}{n^2}$. Thus we obtain an operator which has the same spectrum as the quantum mechanical energy operator on the negative energy states of the hydrogen atom.

REFERENCES

1. R. J. Blattner, *Quantisation and representation theory,* Proc. Sympos. Pure Math. vol. 26, Amer. Math. Soc. Providence R.I. 1974, pp 147-165.

2. B. Kostant, *Quantisation and unitary representations,* Lecture Notes in Math., vol. 170, Springer-Verlag, Berlin, 1970, pp 87-208.

3. D. J. Simms, *Geometric quantisation of energy levels in the Kepler problem,* Conv. Geom. Simpl. Fis. Mat., INDAM, Rome, 1973, to appear in Symposia Math. series, Academic Press.

School of Mathematics
39 Trinity College
DUBLIN
Ireland

Continuation analytique de la série discrète holomorphe

Exposé de Michèle Vergne

sur un travail commun avec Hugo Rossi

1.1 Exposé du problème

Soit G une algèbre de Lie simple et $G = k + P$ une décomposition de Cartan de G; on suppose que le centre z de k est non nul, on a alors $z = RZ$, où les valeurs propres de Z agissant dans P^C sont $\pm i$; on pose

$P^+ = \{X \in P^C;\ \ [Z,X] = i\,X\}$

$P^- = \{X \in P^C;\ \ [Z,X] = -i\,X\}$

et on a $k = [k,k] + RZ$.

Soit $\tilde{G}$ le groupe simplement connexe d'algèbre de Lie G. Si $X \in G$ et si ϕ est une fonction differentiable sur $\tilde{G}$ on note $r(X)\phi$ la fonction définie par $(r(X)\phi)(g) = \frac{d}{dt}\phi(g \exp tX)\big|_{t=0}$, et si $X \in G^C$, on définit $r(X)$ par linéarité.

Soit $\tilde{K}$ le sousgroupe analytique de $\tilde{G}$ d'algèbre de Lie k; alors $\tilde{G}/\tilde{K}$ est un espace hermitien symmétrique, les fonctions holomorphes sur $\tilde{G}/\tilde{K}$, pouvant être identifiées aux fonctions sur $\tilde{G}$ annulées par tous les champs de vecteurs $r(X)$ pour $X \in k^C + P^-$.

Remarquons que $\tilde{K}$ n'est pas compact, on a $\tilde{K} = [\tilde{K},\tilde{K}] \exp RZ$, et $[\tilde{K},\tilde{K}]$ est compact; le centre $Z(\tilde{G})$ de $\tilde{G}$ est isomorphe à $\mathbb{Z}$; soit alors U_0 une représentation unitaire irréductible fixe de $[\tilde{K},\tilde{K}]$ dans un espace vectoriel V_0 (de dimension finie) et λ un caractère (variable) de $\tilde{K}$, alors $U_0 \otimes \lambda = U_\lambda$ est une représentation irréductible de $\tilde{K}$ dans V_0.

1.2 On definit

$\mathcal{O}(U_0;\lambda) = \{\phi$ fonctions C^∞ sur $\tilde{G}$ à valeurs dans V_0 vérifiant $\phi(gk) = U_\lambda(k)^{-1}\cdot\phi(g)$ et $r(X)\cdot\phi = 0$ quelque soit $X \in P^-\}$.

$\mathcal{O}(U_0;\lambda)$ s'identifie à l'espace des sections holomorphes du fibre holomorphe $\tilde{G} \times V_0/\tilde{K} \to \tilde{G}/\tilde{K}$. Le groupe $\tilde{G}$ agit par translations à gauche dans $\mathcal{O}(U_0;\lambda)$ par $(T_{U_0,\lambda}(x)\phi)(g) = \phi(x^{-1}g)$.

1.3 <u>On définit</u>

$$H(U_0;\lambda) = \{\phi \in \mathcal{O}(U_0;\lambda) \text{ et telles que}$$

$$N(U_0;\lambda;\phi)^2 = \int_{\tilde{G}/Z(\tilde{G})} \|\phi\ (g\ )\|^2_{V_0}\ dg < +\infty .$$

Alors si $H(U_0;\lambda)$ est non nul, il est facile de voir que la représentation $T_{U_0,\lambda}$ de $\tilde{G}$ par translations à gauche dans $H(U_0;\lambda)$ est unitaire irréductible et appartient à la série discrète (relative) de $\tilde{G}$ [7]. Comme $H(U_0;\lambda)$ s'identifie à l'espace des sections holomorphes d'un fibre hermitien, de carré intégrable, on dit que la représentation $T_{U_0,\lambda}$ de $\tilde{G}$ dans $H(U_o;\lambda)$ est un membre de la série discrète holomorphe (relative) de $\tilde{G}$.

Harish-Chandra [7] donne une condition nécessaire et suffisante pour que $H(U_0;\lambda) \neq \{0\}$.

1.4 <u>Introduisons quelques notations:</u>

On choisit h une sous algèbre abélienne maximale de k, alors $h^{\mathbb{C}}$ est une sous algèbre de Cartan de $G^{\mathbb{C}}$; l'ensemble Δ des racines de $G^{\mathbb{C}}$ par rapport à $h^{\mathbb{C}}$ est la réunion disjointe de l'ensemble Δ_K des racines compactes (i.e. $G^{\alpha} \subset k^{\mathbb{C}}$) et de l'ensemble Δ_P des racines non compactes; on choisit un ordre total sur les racines, et tel que $P^+ = \sum_{\alpha \in \Delta_P^+} G^{\alpha}$.

Pour chaque $\gamma \in \Delta$, on note H_γ l'unique vecteur de $[G^{\gamma},G^{-\gamma}] \cap ih$ tel que $\gamma(H_\gamma) = 2$, et si $\gamma \in \Delta_P^+$ on choisit $E_\gamma \in G^{\gamma}$ et $E_{-\gamma} = \overline{E_\gamma}$ ($X \to \overline{X}$ désigne la conjugaison dans G^C par rapport à la forme réelle G de G^C) tels que $[E_\gamma,E_{-\gamma}] = H_\gamma$.

Soit $\rho = \frac{1}{2} \sum_{\alpha \in \Delta_P^+} \alpha$ et soit Λ le plus haut poids de la représentation $U_\lambda = U_0 \otimes \lambda$; lorsque le caractère λ varie, U_0 restant fixe la restriction de Λ à $h \cap [k,k]$ reste fixe; on la note Λ_0.

Soit r le rang réel de G et γ_r la plus grande racine, comme $H_{\gamma_r} \notin [k,k]^{\mathbb{C}}$, Λ est entièrement déterminée par Λ_0, forme linéaire sur $h \cap [k,k]$ telle que $\langle\Lambda_0, H_\alpha\rangle$ soit un entier positif ou nul, pour toute racine $\alpha \in \Delta_K^+$, et par $\lambda = \langle\Lambda, H_{\gamma_r}\rangle$ qui par contre est un nombre réel quelconque.

1.5 $\Lambda = (\Lambda_0, \lambda)$ étant donné, on notera U_Λ la représentation de $\tilde{K}$ de plus haut poids Λ dans V_Λ et on notera $\mathcal{O}(U_0;\lambda) = \mathcal{O}(\Lambda)$, $H(U_0;\lambda) = H(\Lambda)$, et $T_{U_0,\lambda} = T_\Lambda$. Harish-Chandra [7] a montré que $H(\Lambda) \neq 0$ si et seulement si $\langle\Lambda + \rho, H_{\gamma_r}\rangle < 0$.

Notons $\ell_r = \langle\rho, H_{\gamma_r}\rangle$, c'est un nombre entier strictement positif; Λ_0 étant fixe, la série discréte holomorphe T_Λ est donc paramètrée par l'ensemble des $\lambda = \langle\Lambda, H_{\gamma_r}\rangle$ strictement infèrieurs à $-\ell_r$

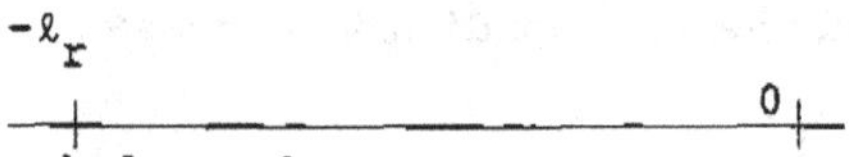

série discrète holomorphe

De manière vague, notre problème est de montrer que la norme $N(\Lambda;\phi)^2 = \int ||\phi||^2 d\dot{g}$ peut se prolonger pour des valeurs $\lambda \geq -\ell_r$, aprés multiplication par un facteur méromorphe convenable $c(\lambda)$ en une norme $N_0(\Lambda;\phi)$ strictement positive sur un sous espace non nul $H_0(\Lambda)$ de $\mathcal{O}(\Lambda)$; on obtiendra ainsi de nouvelle représentations unitaires irréductibles de $\tilde{G}$, qu'on explicitera autant que possible.

Knapp et Okamoto [8] ont montré la possibilité d'une telle construction pour $\lambda = -\ell_r$. K. Gross, R. Kunze et N. Wallach en nous montrant divers exemples frappants de la possibilité de telles constructions pour $\lambda > -\ell_r$ ont tourné notre attention sur ce problème (voir [6] et [16]), et notre travail a été influencé par leurs idées sur ce sujet. Nous n'avons appris les résultats les plus récents de N. Wallach (voir exposé de Wallach dans ce même volume) qu'aprés la conférence de Luminy.

1.6 Enonçons maintenant de maniére plus précise ce problème:

Soit $G_{\mathbb{C}}$ le groupe simplement connexe d'algèbre de Lie $\mathcal{G}^{\mathbb{C}}$, et G, K, $K_{\mathbb{C}}$, P_+ et P_- les sous groupes connexes de $G_{\mathbb{C}}$ d'algèbres de Lie $\mathcal{G}$, k, $k^{\mathbb{C}}$, P^+ et P^-. Observons que $\tilde{G}/\tilde{K}$ est canoniquement isomorphe à G/K.

1.7 Tout élément de $P_+ K_{\mathbb{C}} P_-$ s'écrit de manière unique $g = \exp \zeta(g)\cdot k(g)\cdot \exp \zeta'(g)$, avec $\zeta(g) \in P^+$, $k(g) \in K_{\mathbb{C}}$ et $\zeta'(g) \in P^-$.

On a $G \subset P_+ K_{\mathbb{C}} P_-$, et l'application $g \to k(g)$ se relève en une application qu'on notera encore $k(g)$ de $\tilde{G}$ dans $\tilde{K}_{\mathbb{C}}$, recouvrement universel de $K_{\mathbb{C}}$. D'autre part, on notera encore U_Λ la représentation holomorphe de $\tilde{K}_{\mathbb{C}}$ qui se restreint à $\tilde{K}$ suivant U_Λ.

1.8 On définit $\Phi^0_\Lambda(g) = U_\Lambda(k(g))$

Si ϕ est une fonction holomorphe sur $\tilde{G}/\tilde{K}$ à valeurs dans l'espace vectoriel V_Λ, la fonction $g \to \Phi^0_\Lambda(g)^{-1}\cdot\phi(g)$ est une fonction dans $\mathcal{O}(\Lambda)$ et on obtient ainsi tout élément de $\mathcal{O}(\Lambda)$.

1.9 En particulier, si v est une vecteur de V_Λ, on définit $\Psi^v_\Lambda(g) = \Phi^0_\Lambda(g)^{-1}\cdot v$, c'est une fonction particuliére de $\mathcal{O}(\Lambda)$. Soit v_Λ le vecteur de plus haut poids de la représentation U_Λ (et normalisé de telle sorte que $||v_\Lambda|| = 1$).

1.10 On définit $\Psi_\Lambda(g) = \langle\Psi^{v_\Lambda}_\Lambda(g), v_\Lambda\rangle$.

Ψ_Λ est une fonction sur $\tilde{G}$ à valeurs scalaires.

1.11 On note $L_\Lambda \subset \mathcal{O}(\Lambda)$ le sous espace de $\mathcal{O}(\Lambda)$ engendré par les translatées à gauche de la fonction $\Psi^{v_\Lambda}_\Lambda$.

On voit facilement [7] que si $H(\Lambda) \neq \{0\}$ alors $\Psi^{v_\Lambda}_\Lambda \in H(\Lambda)$ et que

1.12 quelque soit $\phi \in H(\Lambda)$:

$$\langle\phi(e), v_\Lambda\rangle = \frac{1}{\langle\Psi^{v_\Lambda}_\Lambda, \Psi^{v_\Lambda}_\Lambda\rangle} \langle\phi, \Psi^{v_\Lambda}_\Lambda\rangle_{H(\Lambda)}$$

1.13 On note $C(\Lambda) = \langle\Psi^{v_\Lambda}_\Lambda, \Psi^{v_\Lambda}_\Lambda\rangle$.

La relation (1.12) peut encore s'écrire

1.14 $$<\psi_\Lambda^{\vee\Lambda}, T_\Lambda(x)\psi_\Lambda^{\vee\Lambda}> = C(\Lambda)\psi_\Lambda(x) .$$

Par conséquent si $H(\Lambda) \neq 0$, L_Λ est un sous-espace dense de $H(\Lambda)$ et on peut calculer la norme d'un élément de L_Λ grâce a la fonction ψ_Λ , i.e.,

$$||\sum_{i=1}^{N} c_i T(g_i)\psi_\Lambda^{\vee\Lambda}||^2 = C(\Lambda)(\Sigma\, c_i\overline{c_j}\psi_\Lambda(g_i^{-1}g_j)) .$$

La fonction ψ_Λ est donc de type positif lorsque $\lambda < -l_r$.

1.15.<u>Problème A</u>: Λ_0 étant fixé , calculer l'ensemble P_{Λ_0} des nombres réels λ , tel que si Λ est donné par (Λ_0,λ) , ψ_Λ soit de type positif, i.e., $\Sigma c_i\overline{c_j}\psi_\Lambda(g_i^{-1}g_j) \geq 0$.

Il est clair que si $\lambda \in P_{\Lambda_0}$, on peut alors munir L_Λ d'un produit scalaire invariant par les translations à gauche par

$$||\Sigma c_i T(g_i)\psi_\Lambda^{\vee\Lambda}||^2 = \Sigma\, c_i\overline{c_j}\psi_\Lambda(g_i^{-1}g_j) .$$

1.16. On notera $H_0(\Lambda)$ le complété de L_Λ pour la norme ci-dessus, et $N_0(\Lambda;\phi)$ la norme d'un élément ϕ de $H_0(\Lambda)$. Donc si $\lambda < -l_r$, on a $H_0(\Lambda) = H(\Lambda)$, mais $N_0(\Lambda;\phi)^2 = C(\Lambda)^{-1}N(\Lambda;\phi)^2$.

Il est facile de montrer [11]:

1.17. <u>Proposition</u>. Si $\lambda \in P_{\Lambda_0}$ alors $H_0(\Lambda)$ s'identifie à un sous-espace de $\mathcal{O}(\Lambda)$; si $F \in H_0(\Lambda)$ on a $<F(e),v_\Lambda> = <F,\psi_\Lambda^{\vee\Lambda}>_{H_0(\Lambda)}$ et la représentation T_Λ de $\tilde{G}$ par translations à gauche dans $H_0(\Lambda)$ est une représentation unitaire irreductible de $\tilde{G}$.

<u>Probleme B</u>. Identifier autant que possible les représentations unitaire irreductibles T_Λ ainsi obtenues, et expliciter l'espace $H_0(\Lambda)$.

2. Un résultat algèbrique

Si $X \in G$, et si ϕ est une fonction indéfiniment différentiable sur $\tilde{G}$, on pose $X \cdot \phi = \frac{d\phi}{dt}(\exp(-tX)g)\big|_{t=0}$. On prolonge cette action en une action de l'algèbre enveloppante U de G^C, et cette action laisse evidemment $\mathcal{O}(\Lambda)$ stable.

2.1 On note $W_\Lambda = U \cdot \Psi_\Lambda^{v_\Lambda} \subset \mathcal{O}(\Lambda)$. On voit facilement que W_Λ est isomorphe au module de Wallach [15] et est donc irréductible comme G-module. Soit $G^+ = \sum_{\alpha \in \Delta^+} G^\alpha$, on a $X \cdot \Psi_\Lambda^{v_\Lambda} = 0$, quelque soit $X \in G^+$ et $H \cdot \Psi_\Lambda^{v_\Lambda} = \langle \Lambda, H \rangle \Psi_\Lambda^{v_\Lambda}$, de sorte que W_Λ est l'unique quotient simple du module de Verma $M_{\Lambda+\rho}$ de plus haut poids Λ (voir [3], chapitre 7).

Soit a_Λ l'annulateur de v_Λ dans $U(k^C)$ et I_Λ l'idéal à gauche engendré par X_α et $H - \Lambda(H)$. Il est clair que l'annulateur J_Λ de $\Psi_\Lambda^{v_\Lambda}$ dans U contient $I_\Lambda + Ua_\Lambda$.
$\alpha \in \Delta^+$

2.2 Comme nous l'a montré Nicole Conze, on a la Proposition:

Si $\langle \Lambda + \rho, H_\gamma \rangle \notin \{1,2,\ldots,n,\ldots\}$, pour toute racine non compacte positive γ, alors $J_\Lambda = I_\Lambda + Ua_\Lambda$.

On peut voir sur des exemples particuliers que cette condition est cependant loin d'être nécessaire pour que l'idéal $I_\Lambda + Ua_\Lambda$ soit maximal. On en déduit cependant par un argument de continuité le:

2.3 Corollaire

Si $\lambda + \ell_r \leq 1$, alors $\lambda \in P_{\Lambda_0}$.

Ceci nous indique déjà qu'il est possible de dépasser le point limite $\lambda = -\ell_r$ pour la construction des espaces $H_0(\Lambda)$.

2.4 Nous supposerons désormais que $\Lambda_0 = 0$; U_Λ est donc un caractère de $\tilde{K}$; on notera indifféremment λ le nombre $\lambda = \langle \Lambda, H_{\gamma_r} \rangle$, la forme $\Lambda = (0,\lambda)$ sur $h^{\mathbb{C}} = (h \cap [k,k])^{\mathbb{C}} + \mathbb{C}H_{\gamma_r}$, et le caractère de $\tilde{K}$ correspondant. On dira que λ appartient à P si λ appartient à P_0 ($\Lambda_0 = 0$) (def. 1.15), et on notera $\mathcal{O}(\lambda)$, $H_0(\lambda)$ les espaces correspondants.

3. Espaces de Hilbert de fonctions holomorphes sur un domaine de Siegel.

Rappelons la définition d'un domaine de Siegel:

Soit Ω un cône convexe ouvert dans un espace vectoriel réel de dimension finie V, s un point de base de Ω; on considère $\Omega^* = \{\xi \in V^*$, tels que $\langle\xi,y\rangle > 0$ quelque soit $y \in \bar{\Omega} - \{0\}\}$, et on suppose Ω propre, i.e. $\Omega^* \neq \phi$.

On note $D(\Omega)$ le domaine $V + i\Omega \subset V^{\mathbb{C}}$.

Soit W un espace vectoriel complexe et $Q(u,v): W\times W \to V^{\mathbb{C}}$ une forme sesquilinéaire Ω-hermitienne i.e. telle que $Q(u,u) \in \bar{\Omega} - \{0\}$ si $u \neq 0$. Le demi plan de Siegel $D(\Omega:Q)$ est alors le domaine ouvert de $V^{\mathbb{C}} \times W$:

$$D(\Omega,Q) = \{p = (x+iy,u) \in V^{\mathbb{C}} \times W;\ \ y - Q(u,u) \in \Omega\}.$$

On sait [10] que tout espace hermitien symétrique G/K est réalisable sous la forme d'un domaine $D(\Omega;Q)$. Remarquons que si $p_1 = (z_1,u_1)$ et $p_2 = (z_2,u_2)$ sont des points de $D(\Omega;Q)$, alors $z_1 - \bar{z}_2 - 2i\,Q(u_1,u_2) \in D(\Omega)$. Soit $R(p_1,p_2)$ une fonction sur $D(\Omega;Q) \times D(\Omega;Q)$ holomorphe par rapport à p_1, et antiholomorphe par rapport à p_2.

On dira que R est une fonction Noyau si R vérifie la condition

3.1 (P) $\Sigma\lambda_i\bar{\lambda}_j\,R(p_i,p_j) \geq 0$, quelque soit la suite $p_1,p_2,\ldots,p_N$ de points de $D(\Omega;Q)$, et $(\lambda_1,\lambda_2,\ldots,\lambda_N)$ de nombres complexes.

A chaque fonction Noyau R, on peut associer de manière abstraite un espace de Hilbert $H_0(R)$ et un seul de fonctions holomorphes sur $D(\Omega;Q)$ dont R soit le Noyau reproduisant, i.e., tel que $F(p) = \langle F,R(?,p)\rangle$ pour tout $F \in H_0(R)$ (voir par exemple

Kunze [12]). Dans le cas d'un domaine symétrique $D(\Omega;Q) \simeq \tilde{G}/\tilde{K}$ et de notre caractère λ de $\tilde{K}$, nous nous intéresserons aux fonctions R_λ définies de la manière suivante:

3.2 Soit $k(z) = \int_{\Omega^*} e^{2i\pi<\xi,z>}d\xi$ la fonction de Köecher. Cette intégrale converge absolument quelque soit $z \in D(\Omega)$ et définit un fonction holomorphe partout non nulle sur $D(\Omega)$. On normalisera $d\xi$, mesure euclidienne sur V^* de telle sorte que $k(is) = 1$; on peut alors définir les puissances k^α de k pour n'importe quel nombre réel α normalisées par $k^\alpha(is) = 1$.

Et nous aurons à considérer la fonction

$$R_\lambda(P_1,P_2) = k\left(\frac{z_1 - \overline{z}_2 - 2i\, Q(u_1,u_2)}{2}\right)^{\frac{-r\lambda}{n}}$$

où r est le rang de G/K et n la dimension de l'espace vectoriel $V \supset \Omega$. Par exemple: si G est le groupe $SP(r;\mathbb{R})$, G/K est réalisable comme le domaine $D(\Omega)$, où Ω est le cône des matrices $(r \times r)$ symétriques définies positives, et on a:

$$R_\lambda(z,w) = \det\left(\frac{z - \overline{w}}{2i}\right)^{-\lambda}.$$

On montrera que Ψ_λ est de type positif si et seulement si R_λ est une fonction noyau, i.e. vérifie la condition (P) donnée en (3.1), et l'espace $H_0(\lambda)$ sera naturellement isomorphe à l'espace $H_0(R_\lambda)$.

3.3. Donnons maintenant dans le cas d'un domaine de Siegel $D(\Omega;Q)$ quelconque un critère suffisant pour qu'une fonction R vérifie (P) et une description "concrète" de l'espace $H_0(R)$ correspondant.

3.4 Soit J une fonction continue strictement positive sur Ω^*, et supposons J homogène par rapport aux homothéties, i.e. $(J(t\xi) = t^\alpha J(\xi)$ pour α réel quelconque et $t > 0)$. On note $L^2(\Omega^*;Q,J)$ = espace des fonctions mesurables $\ell(\xi,u)$ sur $\Omega^* \times W$ et telle que $||\ell||_J^2 = \int|\ell(\xi,u)|^2 e^{-4\pi<\xi,Q(u,u)>}J(\xi)\, d\xi\, du$ soit finie; (du est la mésure euclidienne de l'espace W considéré comme espace réel) et on considére $\mathcal{H}(\Omega^*;Q;J)$ l'espace des fonctions mesurables $\phi(\xi,u)$ dans $L^2(\Omega^*;Q,J)$ telles que $\phi(\xi,u)$ soit holomorphe en u pour presque tout $\xi \in \Omega^*$. L'ensemble des classes d'équivalence $H(\Omega^*;Q;J)$ de telles fonctions est un sous espace fermé de $L^2(\Omega^*;Q,J)$.

3.5 D'autre part si $\xi \in \Omega^*$, la forme $Q_\xi(u,v) = 4<\xi,Q(u,v)>$ est une forme hermitienne définie positive sur W. On note $\det Q_\xi$ le déterminant de cette forme par rapport à une base complexe choisie une fois pour toutes de W.

3.6 Définition

On dira que J vérifie la condition (C) si

$$G_J(y) = \int_{\Omega^*} e^{-4\pi<\xi,y>} J(\xi)^{-1} \det Q_\xi \, d\xi$$

converge en un point $y_0 \in \Omega$.

Dans le cas où J vérifie (C), alors $G_J(y)$ converge en tout point y de Ω; et on définit

3.7
$$G_J(z) = \int_{\Omega^*} e^{+4i\pi<\xi,z>} J(\xi)^{-1} \det Q_\xi \, d\xi.$$

C'est une fonction holomorphe sur Ω, et on définit

3.8
$$R_J(p_1,p_2) = \int_{\Omega^*} e^{2i\pi<\xi,z_1-\bar{z}_2-2iQ(u_1,u_2)>} J(\xi)^{-1} (\det Q_\xi) d\xi$$

où $p_1 = (z_1,u_1)$, $p_2 = (z_2,u_2)$.

La Proposition suivante montre que si J vérifie la condition (C) alors R_J vérifie la condition (P), et donne une description de $H_0(R_J)$ à l'aide de la transformation de Fourier-Laplace.

3.9 Proposition

Supposons que J vérifie la condition (C), alors quelle que soit $\phi \in \mathcal{H}(\Omega^*;Q,J)$

$$\check{\phi}(z,u) = \int_{\Omega^*} e^{2i\pi<\xi,z>} \phi(\xi,u) d\xi$$

converge absolument en tout point $p = (z,u)$ de $D(\Omega;Q)$ et uniformément sur les compacts.

On a $|\check{\phi}(z,u)|^2 \leq G(y - Q(u,u))||\phi||_J^2$.

On notera $\check{H}(\Omega^*;Q,J)$ l'espace des fonctions $\check{\phi}(z,u)$ pour $\phi \in H(\Omega^*;Q,J)$ muni de la norme $||\check{\phi}|| = ||\phi||_J$.

Alors $\check{H}(\Omega^*;Q,J)$ est un espace de Hilbert de fonctions holomorphes sur $D(\Omega;Q)$ ayant comme noyau reproduisant R_J.

3.10 Proposition [14]

Soit Ψ une fontion homogène strictement positive sur le cône Ω telle que $I_\Psi(\xi) = \int_\Omega e^{-4\pi<\xi,y>}\Psi(y)\,dy$ converge en un point $\xi \in \Omega^*$ (et donc pout tout point $\xi \in \Omega^*$), alors $J = I_\Psi$ vérifie la condition (C), et on a $\check{H}(\Omega^*, Q, I_\Psi) = \{F$ hol. sur $D(\Omega;Q)$ telles que $||F||^2 = \int|F(x + iy,u|^2\Psi(y - Q(u,u))dx\,dy\,du < +\infty\}$. Réciproquement ce dernier espace est $\neq \{0\}$ si et seulement si I_Ψ converge en un point $\xi \in \Omega^*$.

On introduit alors la condition suivante:

3.11 Définition: On dit que J vérifie la condition d'Harish-Chandra, notée (H.C.), si $J = I_\Psi$ pour une fonction Ψ sur Ω.

La condition (H.C.) entraine donc la condition (C). Cependant cette condition (H.C.) est certainement plus restrictive que la condition (C); par exemple, la fonction $J_0 = 1$ vérifie la condition (C) et dans ce cas, l'espace

$\check{H}(\Omega^*;Q,1) = H^2 = \{F$ hol. sur $D(\Omega;Q)$ telles que

$$\sup_{t\in\Omega}\int|F(x + i(t + Q(u,u)),u)|^2dx\,du = ||F||^2 < +\infty\}.$$

C'est-à-dire, c'est l'espace de Hardy [9] mais J_o n'est pas de la forme I_Ψ.

Dans le cas d'un cône Ω homogène sous l'action d'un sous groupe H de $GL(V)$ et d'une fonction J vérifiant $J(h\cdot\xi) = y(h)J(\xi)$ pour y caractère de H, nous pouvons donner des critères numériques nécessaires et suffisants sur y pour J vérifie la condition (H.C) ou la condition (C).

4 - Réalisation des espaces $H_0(\lambda)$.

Nous allons expliciter dans le cas du domaine de Siegel symétrique $\tilde{G}/\tilde{K}$ la méthode précédente. On reprend les notations du paragraphe 1.

4.1 Soit $\Psi = (\gamma_1, \gamma_2, \ldots, \gamma_r)$ l'ensemble des racines non compactes positives choisies de la manière suivante: γ_r est la plus grande racine et γ_i est la plus grande racine orthogonale à $(\gamma_{i+1}, \ldots, \gamma_r)$. On pose $X_\gamma = E_\gamma + E_{-\gamma}$. Alors $a = \sum_{\gamma \in \Psi} RX_\gamma$ est une sous algèbre abélienne maximale de P. On définit si $\gamma \in \Delta_p^+$ l'élément c_γ de $G_{\mathbb{C}}$ par

4.2 $c_\gamma = \exp - \frac{\pi}{4}(E_\gamma - E_{-\gamma})$

et on définit la transformation de Cayley c par

4.3 $c = \prod_{i=1}^{r} c_{\gamma_i}$.

On a $c(H_{\gamma_i}) = X_{\gamma_i}$.

4.4 On note $(\alpha_1, \alpha_2, \ldots, \alpha_r)$ les restrictions à a des transformées par c des racines $(\gamma_1, \gamma_2, \ldots, \gamma_r)$. On a donc $\alpha_i(X_{\gamma_j}) = 2\delta_i^{\,j}$.

On choisit un ordre sur l'espace vectoriel dual a^* de a tel que $\alpha_1 < \alpha_2 < \ldots < \alpha_r$.

On a alors

4.5 Théorème (C.C. Moore) [13].

Toute racine $\alpha > 0$ de G par rapport à a est de la forme

$$\tfrac{1}{2}(\alpha_i + \alpha_j), \quad 1 \leq i \leq j \leq r \qquad \tfrac{1}{2}(\alpha_i - \alpha_j), \quad 1 \leq j < i \leq r \qquad \text{ou} \quad \tfrac{1}{2}\alpha_i .$$

La multiplicité de la racine α_i est 1.

La multiplicité de $\frac{\alpha_i + \alpha_j}{2}$ pour $i \neq j$ est egale à la multiplicité de $\frac{\alpha_i - \alpha_j}{2}$ et est indépendante du couple (i,j). La multiplicité de $\frac{\alpha_i}{2}$ est paire et indépendante de i.

4.6 On posera p = mult. de

$\frac{\alpha_i + \alpha_j}{2}$, $i \neq j$, et μ = mult. de $\frac{\alpha_i}{2}$. On a alors $\langle \rho, H_{\gamma_r} \rangle = \ell_r = (r-1)p + 1 + \frac{\mu}{2}$. Enonçons maintenant notre résultat concernant la Problème (A), dans le cas $\Lambda_0 = 0$.

4.7 Théorème

Si $\lambda \leq \frac{(r-1)p}{2}$, alors Ψ_λ est de type positif.

4.8 On notera $\ell_0 = \frac{(r-1)p}{2}$. Pour démontrer ce théorème, nous allons relier la fonction Ψ_λ avec la fonction R_λ (3.2) et on exprimera R_λ lorsque $\lambda \leq (r-1)\frac{p}{2}$ comme une transformée de Fourier-Laplace R_{J_λ}.

Rappellons la réalisation de $\tilde{G}/\tilde{K}$ comme un domaine $D(\Omega;Q)$ [10].

5.1 Soit $G = k \oplus a \oplus n$ la décomposition d'Iwasawa de G. On notera $b = a \oplus n$ et $b^+ = b^{\mathbb{C}} \cap (k^{\mathbb{C}} \oplus P^+)$ et $b^- = b^{\mathbb{C}} \cap (k^{\mathbb{C}} \oplus P^-)$ et soit J l'endomorphisme réel de b, qui vaut $+i$ sur b^+ et $-i$ sur b^-. On a $J(X_{\gamma_i}) \in n^{\alpha_i}$,

$$J(n^{\frac{\alpha_i-\alpha_j}{2}}) = n^{\frac{\alpha_i+\alpha_j}{2}} \quad \text{et} \quad J(n^{\frac{\alpha_i}{2}}) = n^{\frac{\alpha_i}{2}}, \quad i > j.$$

5.2 On note $U_i = -\frac{1}{2} JX_{\gamma_i}$ et

5.3 $s = \sum_{i=1}^{r} U_i$.

L'élément $Js = \frac{1}{2} \sum_{i=1}^{r} X_{\gamma_i}$ est un élément de a qui a comme valeurs propres 0, ± 1 et éventuellement $\pm\frac{1}{2}$.

On note $b = H_0 \oplus H_{1/2} \oplus H_1$, la décomposition de b en sous espaces propres par rapport à ad Js; i.e. $H_0 = a + \sum_{i>j} n^{\frac{\alpha_i-\alpha_j}{2}}$, $H_{1/2} = \sum_i n^{\frac{\alpha_i}{2}}$, $H_1 = \sum_{i \leq j} n^{\frac{\alpha_i+\alpha_j}{2}}$. On note B, A, N, H_0 les sous groupes analytiques de $\tilde{G}$ d'algèbre de Lie b, a, n, H_0 et tout élément de B s'écrit de manière unique $b = h_0 \cdot \exp U \cdot \exp X$ avec

$h_0 \in H_0$, $U \in H_{1/2}$, $X \in H_1$. Le groupe H_0 agit dans $H_{1/2}$ et H_1 par l'action adjointe.

5.4 L'orbite Ω de $s \in H_1$ par le groupe H_o est un cône convexe propre de H_1; l'élément ξ_0 de $H_1{}^*$ défini par

$$\xi_0\Big(\sum_{i<j} n^{\frac{\alpha_i+\alpha_j}{2}}\Big) = 0 \text{ et } \xi_0(U_i) = 1, \text{ est un élément de } \Omega^*.$$

L'application $h \to h\cdot s$ (resp. $h \to h\cdot \xi_0$) est un difféomorphisme de H_0 sur Ω (resp. sur Ω^*). On note $h(t)$ (resp. $h(\xi)$) l'application inverse.

5.5 L'espace $H_{1/2}$ est stable par J; on note $H_{1/2}^+ = (H_{1/2})^{\mathbb{C}} \cap b^+$; l'application $\tau(u) = \frac{1}{2}(u - iJu)$ envoie $H_{1/2}$ sur $H_{1/2}^+$. Le groupe H_0 laisse stable $H_{1/2}^+$.

5.6 La forme $Q(u,v) = \frac{i}{2}([u,\bar{v}])$ est une forme Ω-hermitienne, définie sur $H_{1/2}^+$. On peut donc définir le domaine $D(\Omega;Q) \subset H_1{}^{\mathbb{C}} \oplus H_{1/2}^+$. La transformation de Cayley c (agissant dans G^C par l'action adjointe transforme P^+ en $H_1{}^{\mathbb{C}} \oplus H_{1/2}^+$.

5.7 Considérons l'application ζ de G dans P^+; cette application réalise un isomorphisme de G/K sur un ouvert borné D de P^+; on sait d'autre part que $c^{-1}G \subset P_+K_{\mathbb{C}}P_-$. On peut donc définir

$$\alpha(g) = c\cdot\zeta(c^{-1}g) \in H_1{}^{\mathbb{C}} \oplus H_{1/2}^+.$$

On a, si $b = h_0\cdot \exp U\cdot \exp X$,

5.8 $\alpha(h_0\cdot \exp U\cdot \exp X) = (h_0\cdot X + i\, h_0\cdot(s + Q(\tau(u),\tau(u))), h_0\tau(u))$ et α réalise un isomorphisme de $\tilde{G}/\tilde{K}$ sur $D(\Omega;Q)$. On a $\alpha(1) = p_0 = (is,0)$ et on a $\alpha(b) = b\cdot(is,0)$. D'autre part les translations à gauches par le groupe B sont des automorphismes affines de l'espace $H_1{}^{\mathbb{C}} \oplus H_{1/2}^+$, on note $b\cdot p$ l'action de B sur $H_1{}^{\mathbb{C}} \oplus H_{1/2}^+$. On dira que G/K est du type I ou tubulaire, si $H_{1/2} = 0$, dans ce cas G/K est réalisable comme $D(\Omega) = V + i\Omega$. Sinon on dira que G/K est du type II.

Comme $c^{-1}G \subset P_+K_{\mathbb{C}}P_-$, on peut définir:

5.9 $\Phi_\lambda(g) = \Phi^0_\Lambda(c^{-1})^{-1}\cdot\Phi^0_\lambda(c^{-1}g)$ où Φ^0_λ est donné par (1.8) et la restriction de Φ_λ à B est un caractère de B. On a précisement:

$$\Phi_\lambda(\prod_{i=1}^{r} \exp a_i X_{\gamma_i} \cdot n) = \prod_{i=1}^{r} e^{a_i\lambda},$$

où $a_i \in \mathbb{R}$, et $n \in N$.

5.10 Si on note $\mathcal{O}(D)$ l'ensemble des fonctions holomorphes sur $D = D(\Omega;Q) \simeq \tilde{G}/\tilde{K}$ alors $(P_\lambda F)(g) = \Phi_\lambda(g)^{-1} F(\alpha(g))$ est un isomorphisme de $\mathcal{O}(D)$ sur $\mathcal{O}(\lambda)$ ($\mathcal{O}(\lambda)$ définie en 1.2 et 2.3; on a $\Psi_\lambda^{\check{\lambda}} = \Psi_\lambda$ dans le cas où $\Lambda_0 = 0$).

Posons $\mu_\lambda(h\cdot s) = \Phi_\lambda(h)^{-2}(\det h_{H_{1/2}})^{-1}(\det h_{H_1})^{-2}$ pour $h \in H_0$. Alors P_λ réalise un isomorphisme unitaire entre $H(\mu_\lambda) = \{F \in \mathcal{O}(D),$ et $N(\lambda;F)^2 = \int|F(x + iy,u)|^2\mu_\lambda(y - Q(u,u))dx\, dy\, du < +\infty\}$ et $H(\lambda)$.

Soit alors

5.11 $I_\lambda(\xi) = \int_\Omega e^{-4\pi<\xi,t>}\mu_\lambda(t)dt$

On a $I_\lambda(\xi) = \Phi_\lambda(h)^{-2}(\det h_{H_1})^{-1}(\det h_{H_{1/2}})^{-1}I_\lambda(\xi_0)$. Si $h.\xi_o = \xi$

On peut facilement calculer $I_\lambda(\xi_0)$ en ramenant l'intégrale sur le cône Ω en une intégrale sur le groupe triangulaire H_0.

5.12 On note $\ell_1 = \frac{(r-1)p}{2} + 1 + \frac{\mu}{2}$ et $\underset{1\leq i\leq r}{\ell_i} = \ell_1 + \frac{(i-1)p}{2}$ (de sorte que $\ell_r = <\rho,H_{\gamma_r}>$). On rappelle que $\ell_0 = \frac{(r-1)p}{2}$. Alors

5.13 $I_\lambda(\xi_0) = \prod_{i=1}^{r}(4\pi)^{\lambda+\ell_i}\Gamma(-(\lambda + \ell_i))$ et l'intégrale définissant $I_\lambda(\xi)$ ne converge que si $\lambda + \ell_r < 0$.

On voit donc que $H(\mu_\lambda)$ et par consequent $H(\lambda) \neq 0$ si et seulement si $\lambda + \ell_r < 0$ (Th. 3.10).

5.14 D'autre part définissons la fonction J_λ sur Ω^* par $J_\lambda(h\cdot\xi_0) = \Phi_\lambda(h)^{-2}(\det h_{H_1})^{-1}(\det h_{H_{1/2}})^{-1}$ de sorte que si $\lambda + \ell_r < 0$, $I_\lambda(\xi) = I_\lambda(\xi_0)J_\lambda(\xi)$.

On voit donc que J_λ vérifie la condition (H.C.) (voir 3.11);

i.e. $H(\mu_\lambda) \neq 0)$, si et seulement si $\lambda + \ell_r < 0$.

5.15 Théorème:

J_λ vérifie la condition (C), si et seulement si $\lambda + \ell_0 < 0$.

Ce théorème s'obtient encore en calculant explicitement J_λ comme intégrale sur le groupe triangulaire H_0.

On introduit enfin comme décrit en 3.2, la fonction $k(z)$ et la fonction R_λ sur $D(\Omega;Q) \times D(\Omega;Q)$;

$$R_\lambda((z_1,u_1),(z_2,u_2)) = k(z_1 - \bar{z}_2 - 2iQ(u_1,u_2))^{-\frac{\lambda r}{n}}.$$

Il est facile de montrer les:

5.16 Proposition

$P_\lambda(R_\lambda(?,p_0)) = \Psi_\lambda$, et la fonction Ψ_λ est de type positif (i.e., λ vérifie (P)) si et seulement si R_λ est une fonction noyau (i.e. R_λ vérifie (P)).

5.16 Proposition

Si $\lambda + \ell_0 < 0$; R_λ est proportionnel à

$$\int_{\Omega^*} e^{2i\pi<\xi, z_1 - \bar{z}_2 - 2iQ(u_1,u_2)>} (\det Q_\xi) J_\lambda(\xi)^{-1} d\xi$$

de sorte que R_λ vérifie (P) si $\lambda + \ell_0 < 0$.

En rassemblant les propositions précédentes, et les résultats du paragraphe 3, on voit que si $\lambda + \ell_0 \leq 0$, alors Ψ_λ est de type positif, et que si $\lambda + \ell_0 < 0$, l'espace $H_0(\lambda)$ est isomorphe via la transformation P_λ à l'espace des fonctions holomorphes $\check{\phi}(z,u)$ sur $D(\Omega;Q)$ obtenues de la manière suivante:

$$\check{\phi}(z,u) = \int_{\Omega^*} e^{2i\pi<\xi,z>} \phi(\xi,u) d\xi$$

où $\phi(\xi,u)$ est une fonction mesurable sur $\Omega^* \times H^+_{1/2}$ qui pour presque tout ξ est holomorphe en u, et telle que

$$||\check{\phi}||^2_J = \int_{\Omega^*} e^{-4\pi<\xi,Q(u,u)>} |\phi(\xi,u)|^2 J_\lambda(\xi) d\xi du$$

soit finie.

L'intégrale définissant $\check{\phi}$ est absolument et uniformement convergente sur tout compact, et $N_0(\lambda.P_\lambda\check{\phi})$ est proportionnelle à $||\check{\phi}||_J$.

6 - Espaces $H_0(R_\lambda)$ et valeurs au bord de fonctions holomorphes.

Considérons la réalisation de G/K comme un domaine borné D de P^+ par l'application ζ (voir 5.7). Alors l'action de G sur D se prolonge continument à l'adhérence $\overline{D}$ de D dans P^+. Le bord topologique $\overline{D} - D$ de D est la réunion de r orbites sous l'action du groupe G[18] ou [17]

6.1 On a la description suivante de ces orbites: on considére les transformations de Cayley partielles $c_e = c_{\gamma_e} \cdot c_{\gamma_{e+1}} \cdot \ldots \cdot c_{\gamma_r}$, qui sont des éléments de $G_{\mathbb{C}}$ contenus dans $P_+K_{\mathbb{C}}P_-$. Alors $\zeta(c_e) \in \overline{D}$ et $\zeta(c_e) \in G\cdot\zeta(c_{e+1})$. Notons $O_e = G\cdot\zeta(c_e)$. On a $\overline{D} - D = \bigcup_{i=1}^{r} O_e$. On a $c_1 = c$ et $G\cdot\zeta(c)$ est le bord de Shilov du domaine D[18]

Considérons l'application $\alpha : G/K \simeq D$ sur $D(\Omega;Q) \subset H_1^{\mathbb{C}} \oplus H_{1/2}^+$. On notera souvent $D(\Omega;Q)$ simplement par D. Pour chaque indice e $1 \leq e \leq r$, α se prolonge continument à $B\cdot\zeta(c_e)$, et le complémentaire de $B\cdot\zeta(c_e)$ dans O_e est de mesure nulle pour l'unique classe de mesure quasi-invariante par G sur O_e.

6.2 Dans les notations du paragraphe 5, on a $\alpha(\zeta(c_e)) = (is_e,0)$ où $s_e = \sum_{i=1}^{e-1} U_i$ appartient à l'adhérence du cône Ω ; en particulier pour $e = 1$, $s_1 = 0$.

6.3 On note $\Sigma_e = B\cdot(is_e,0)$; Σ_e est contenue dans le bord topologique de D.

Si $b = h_0\cdot\exp X\cdot\exp U$, alors

$$b\cdot(is_e,0) = h_0\cdot X + i(h_0\cdot s_e + h_0\cdot Q(\tau(u),\tau(u)),h_0\cdot\tau(u)).$$

On note $O_e = H_0\, s_e$, O_e est contenue dans le bord topologique de Ω, et on voit que $\Sigma_e = \{(x + iy,u)$ avec $y - Q(u,u) \in O_e\}$.

6.4 En particulier $\Sigma_1 = \Sigma = \{(x + iy,u)$ avec $y - Q(u,u) = 0\}$ est le bord de Shilov du domaine $D(\Omega;Q)$.

6.5 Soient $\ell_1, \ell_2, \ldots, \ell_r$ les nombres donnés en (5.12). Nous allons montrer que pour chacune des valeurs $\lambda_e = -\ell_e$, $1 \leq e \leq r$, il est possible d'expliciter la norme d'une fonction $F \in H_0(R_{\lambda_e})$ comme intégrale de la valeur au bord de F sur Σ_e.

6.6 Par exemple, si $e = 1$, i.e. $\lambda_1 = -\ell_1$, on voit que $J_{\lambda_1}(\xi) = 1$, et par conséquent $H_0(R_{\lambda_1})$ = l'espace de Hardy = $H^2 = \{F \in \mathcal{O}(D); \sup \tau \in \Omega \int |F(x + i(\tau + Q(u,u)),u)|^2 dx\, du$
$= ||F||^2 < +\infty\}$.

On note si $\tau \in \Omega$, Σ_τ la surface de niveau τ dans $D(\Omega;Q)$, i.e., $\Sigma_\tau = \{(x + i(\tau + Q(u,u)),u)\} = \Sigma + (i\tau,0)$. Si $F \in H^2$ et $\tau \in \Omega$, la restriction de F à Σ_τ définit une fonction F_τ sur Σ, i.e. $F_\tau(x + iQ(u,u),u) = F(x + i(\tau + Q(u,u)),u))$ et on sait que lorsque $\tau \to 0$, F_τ tend vers une limite $\tilde{F}$ dans $L^2(\Sigma, dx\, du)$ et que l'application $F \to \tilde{F}$ est une isométrie de H^2 sur un sous espace propre de $L^2(\Sigma)$ (Koranyi-Stein) [9].

6.7 On notera $N(Q)$ le groupe $\exp(H_{1/2} \oplus H_1)$; $N(Q)$ agit simplement transitivement sur Σ, et les surfaces Σ_τ sont les orbites de $N(Q)$ dans $D(\Omega;Q)$.

En coordonnées $n = \exp u \cdot \exp x$, on a $n \cdot (i\tau,0) = (x + i(\tau + Q(\tau u,\tau u)),\ \tau u))$; et la mesure de Haar dn s'écrit $dx\, du$.

Donc l'injection précédente envoie $H_0(R_{\lambda_1})$ isométriquement sur un sous espace propre de $L^2(N(Q))$ [9].

6.8 De même on montrera que pour chacun des points $\lambda_e = -\ell_e$, il est possible de considérer $H_0(R_{\lambda_e})$ comme un espace de Hardy partiel.

Pour chaque indice e, on va décrire une fibration π_e de D sur un domaine plus petit D_e. A cette fibration sera associée une représentation du domaine D comme un domaine de Siegel de type III.

$D = \{(p_1, z', u'),\ p_1 \in D_e$ et $\mathrm{Im}\, z' - L_{p_1}(u',u') \in \Omega'\}$ où L_{p_1} est la partie réelle une forme semi-hermitienne dépendant différentiablement de p_1 et Ω' un cône propre d'un sous espace vectoriel $V' \subset H_1$. L'application de fibration étant $\pi_e(p_1, z', u') = p_1$. La fibre au dessus du point $p_1 \in D_e$ est isomorphe à un domaine de Siegel de type II fixe D', et a comme bord de Shilov $\Sigma_{p_1} = \{(p_1, z', u'),\ \mathrm{Im}\, z' = L_{p_1}(u',u')\}$. On aura $\Sigma_e = \bigcup_{p_1 \in D_e} \Sigma_{p_1}$ $\{(p_1, z', u');\ p_1 \in D_e$ et $\mathrm{Im}\, z' = L_{p_1}(u',u')\}$. On considére la surface de niveau $\tau' \in \Omega'$ dans D

$$\Sigma_{e,\tau'} = \{(p_1, a', u'),\ \mathrm{Im}\, z' - L_{p_1}(u',u') = \tau'\} = \bigcup_{p_1 \in D_e} \Sigma_{p_1,\tau'}.$$

On montrera que la restriction $F_{\tau'}$ de F à $\Sigma_{e,\tau'}$ tend vers une limite $\tilde{F}$ sur Σ_e lorsque $\tau' \in \Omega' \to 0$ pour une norme L^2 appropriée sur Σ_e.

Ceci fournira une isométrie de $H_0(R_{\lambda_e})$ sur un sous espace de $L^2(\Sigma_e, d\mu_e)$. On cherchera à caratériser le sous espace en question de fonctions sur Σ_e par les équations de Cauchy-Riemann tangentielles sur Σ_e [5].

Enfin on montrera que l'application $F \to \tilde{F}$ est un opérateur d'entrelacement de la représentation T_{λ_e} avec une représentation d'une série principale induite par une représentation unitaire irréductible d'un sous groupe parabolique maximal P_e de $\tilde{G}$.

6.9 Fibration de D et réalisation de D comme un domaine de Siegel de type III.

Soit $C_e = \{1, 2, \ldots, e-1\}$, et soit D_e le domaine de Siegel construit en considérant le sous ensemble des $(e-1)$ premières racines fondamentales $(\gamma_1, \gamma_2, \ldots, \gamma_{e-1})$. Plus précisement soit

$$H_0(C_e) = \sum_{i \in C_e} RX_{\gamma_i} \oplus \sum_{\substack{i \text{ et } j \in C_e \\ i > j}} n^{\frac{\alpha_i - \alpha_j}{2}}$$

$$H_1(C_e) = \sum_{i,j\varepsilon C_e} n^{\frac{\alpha_i+\alpha_j}{2}}$$

$$H_{1/2}(C_e) = \sum_{i\varepsilon C_e} n^{\frac{\alpha_i}{2}}$$

$$s_e = \sum_{i\varepsilon C_e} U_i$$

et soit $\Omega_e = H_0(C_e)\cdot s_e$, Q_e la restriction de Q à $H_{1/2}(C_e)^+$ et $D_e = D(\Omega_e;Q_e) \subset H_1(C_e)^{\mathbb{C}} \oplus H_{1/2}(C_e)^+$

$$= \{(z_e,u_e);\ \mathrm{Im}\, z_e - Q_e(u_e,u_e) \ \varepsilon\ \Omega_e\}.$$

On considére d'autre part $C'_e = \{e,\ldots,r\}$, $H_0(C'_e)$, $H_1(C'_e) = \sum_{i,j\varepsilon C'_e} n^{\frac{\alpha_i+\alpha_j}{2}}$, $H_{1/2}(C'_e)^+$, et $\Omega'_e = H_0(C'_e)\cdot s'_e$. Enfin on note $H'_{1,e} = \sum_{i\varepsilon C'_e, j\varepsilon C'_e} n^{\frac{\alpha_i+\alpha_j}{2}}$.

Comme dans la suite l'indice e va rester fixe, pour alléger les notations, on l'omettra parfois lorsqu'il n'y aura pas de possibilité de confusion.

Par exemple H'_1 désignera $H'_{1,e}$. On a donc

$$H_1 = H_1(C_e) \oplus H_1(C'_e) \oplus H'_1$$

$$H^+_{1/2} = H_{1/2}(C_e)^+ \oplus H_{1/2}(C'_e)^+.$$

Soit $p = (z,u)$ un point de $D(\Omega;Q) \subset H_1^{\mathbb{C}} \oplus H^+_{1/2}$. Ecrivons $p = (z_1 + z_2 + z',u_1 + u_2)$ avec $z_1 \ \varepsilon\ H_1(C_e)^{\mathbb{C}}$, $z_2 \ \varepsilon\ H_1(C'_e)^{\mathbb{C}}$, $z' \ \varepsilon\ H'^{\mathbb{C}}_1$, $u_1 \ \varepsilon\ H_{1/2}(C_e)^+$, $u_2 \ \varepsilon\ H_{1/2}(C'_e)^+$. Alors $\pi_e(z_1 + z_2 + z',u_1 + u_2) = (z_1,u_1)$ est une application surjective de D dans D_e. (Par exemple si Ω est le cône des matrices symétriques définies positives, on a si

$$z = \left(\begin{array}{c|c} z_1 & z' \\ \hline z' & z_2 \end{array}\right)$$

avec z_1 matrice carrée $(e-1) \times (e-1)$, $\pi_e(z) = z_1$.)

On voit que la fibre au dessus du point $p_1 = (x_1 + iy_1, u_1)$ est l'ensemble des points $(z_1 + z_2 + z', u_1 + u_2)$ avec $\operatorname{Im} z_2 - L_{p_1}(z' + u_2, z' + u_2) \in \Omega'_e$ où $L_{p_1}(z' + u_2, z' + u_2) = \frac{1}{2}[J(h_1^{-1}y'), h_1^{-1}y'] + Q(u_2, u_2)$ si $z' = x' + iy'$, $x', y' \in \mathcal{H}'_1$ et h_1 étant défini par $h_1 \cdot s_1 = y_1 - Q(u_1, u_1)$ et agit sur $\mathcal{H}'_1$ par la représentation adjointe.

6.10. On a $\Sigma_e = \{(z_1 + z_2 + z', u_1 + u_2)$ tels que $(z_1, u_1) \in D_e$ et $\operatorname{Im} z_2 = L_{p_1}(z' + u_2, z' + u_2)\}$. $\Sigma_e \cap \pi_e^{-1}(p_1)$ est le bord de Shilov de $\pi^{-1}(p_1)$. On considére $\tau' \in \Omega'$, et on forme

$\Sigma_{e,\tau'} = \Sigma_e + (i\tau', 0) \subset D(\Omega; Q)$. On voit que Σ_e est naturellement paramétrée par $D_e \times \mathcal{H}_1(C'_e) \times \mathcal{H}_1'^{\mathbb{C}} \times \mathcal{H}_{1/2}(C'_e)^+$.

6.11. On considére la fonction μ_e sur le cône $\Omega_e = H_0(C_e) \cdot s_e$ définie par $\mu_e(h \cdot s_e) = \phi_{\lambda_e}(h)^{-2} (\det_{\mathcal{H}_{1/2}(C_e)} h)^{-1} (\det_{\mathcal{H}_1(C_e)} h)^{-2}$ (voir 5.10)

et $\mu_e^1(h \cdot s_e) = \phi_{\lambda_e}(h)^{-1} (\det_{\mathcal{H}_{1/2}} h)^{-1} (\det_{\mathcal{H}_1} h)^{-2}$. On considére la mesure $d\mu_1$ sur le domaine D_e définie par $d\mu_1 = \mu_e^1(y_1 - Q(u_1, u_1))dx_1 dy_1 du_1$ et la mesure $d\sigma_e$ sur Σ_e produit de $d\mu_1$ par la mesure euclidienne de l'espace vectoriel $\mathcal{H}_1(C'_e) \oplus \mathcal{H}_1'^{\mathbb{C}} \oplus \mathcal{H}_{1/2}(C'_e)^+$.

On définit

$$H_1^e = \{F \in \mathcal{O}(D;\Omega) \text{ telles que}$$

$$\sup_{\tau' \in \Omega'} \int_{\Sigma_e} |F(\sigma + (i\tau', 0)|^2 \, d\sigma < +\infty\}.$$

On note $\bar{\partial}_b$ l'opérateur de Cauchy Riemann induit sur Σ_e [5] et $H_2^e = \{F$ mesurables sur Σ_e, telles que $\int_{\Sigma_e} |F(\sigma)|^2 d\sigma < +\infty$ et telles que $\bar{\partial}_b f = 0$ c.a.d. vérifiant les équations de Cauchy-Riemann tangentielles sur Σ_e au sens distribution$\}$. On montre que si $e \neq 1$, ou si $e = 1$ et G/K n'est pas de type I, alors toute fonction dans H_2^e s'étend en une fonction holomorphe sur D.

6.11 Théorème

On a $H_0(R_{\lambda_e}) = H_1^e$.

Pour tout $F \in H_0(R_{\lambda_e})$, la fonction $F_{\tau'}(\sigma) = F(\sigma + (i\tau',0))$ tend vers une fonction $v_e(F)$ dans $L^2(\Sigma_e, d\sigma)$ et la correspondance $F \to \tilde{F}$ est une isométrie de F dans H_2^e. Si $e \neq 1$, ou si $e = 1$ et G/K n'est pas de type I, alors $F \to v_e(F)$ est un isomorphisme de H_1^e sur H_2^e.

6.12 Maintenant, décrivons la structure CR tangentielle de Σ_e avec une foliation transverse aux fibres de $\pi_e : \Sigma_e \to D_e$. Les feuilles de cette foliation sont les composantes holomorphes par arc de Σ_e (voir Koranyi-Wolf [18]).

Soit $b_e = H_0(C_e) \oplus H_{1/2}(C_e) \oplus H_1(C_e)$. L'application $b_1 \to \alpha_e(b_1) = b_1 \cdot (is_e, 0)$ réalisé un isomorphisme de B_e sur D_e.

Soit $H_0' = \sum_{\substack{i \in C_e' \\ j \in C_e}} n_{\frac{\alpha_i - \alpha_j}{2}}$ et soit $b_e' = H_0(C_e') \oplus H_0' \oplus H_{1/2}(C_e') \oplus H_1' \oplus H_1(C_e')$ et soit $H_{1/2}' = H_0' \oplus H_{1/2}(C_e') \oplus H_1'$. $H_{1/2}'$ est stable par J et comme $JH_1' = H_0'$, on a $H_{1/2}'^{+}$ isomorphe à $H_{1/2}(C_e')^{+} \oplus H_1'^{\mathbb{C}}$. On considère Q' la forme Ω_e'-hermitienne sur $H_{1/2}'^{+}$ définie par (5.6) et $D'(\Omega_e', Q') = D' \subset H_1(C_e')^{\mathbb{C}} \oplus H_{1/2}(C_e')^{+} \oplus H_1'^{\mathbb{C}} = (z_2, u_2 + z')$ avec $\mathrm{Im}\, z_2 - Q'(u_2 + z', u_2 + z') \in \Omega_e\}$. On note $N'(Q'_e) = \exp(H_{1/2}' \oplus H_1(C_e'))$, et Σ' le bord de Shilov de D'. L'application $\alpha_e'(b') = b' \cdot (is_e', 0)$ réalise un isomorphisme de B_e' sur D' et le groupe $N'(Q_e')$ agit simplement transitivement sur Σ' par $n' \cdot (0,0)$. On identifie N' et Σ'. b_e' est une idéal de b et $b = b_e \oplus b_e'$. On note $b_e'^{-} = b_e'^{\mathbb{C}} \cap b^{-}$. Tout élément b de B s'écrit de manière unique: $b = b' \cdot b_1$ avec $b' \in B_e'$ et $b_1 \in B_e$ et $\pi_e(\alpha(b)) = \alpha_e(b_1)$. Donc, si $p_1 = (z_1, u_1) = \alpha_e(b_1)$, on obtient un diffeomorphisme $i_{p_1}: D' \to \pi_e^{-1}(p_1)$ par

$$i_{p_1}(\alpha_e'(b')) = \alpha(b' \cdot b_1) = b' \cdot (z_1, u_1)$$

et i_{p_1} s'étend à $\overline{D'}$, i_{p_1} n'est pas holomorphe.

Par contre, si p' est fixé dans D', et puis, par continuité, si $p' \in \overline{D'}$, l'application $p_1 \to i_{p_1}(p')$ est holomorphe en p_1. C'est-à-dire que l'application $i: D_e \times \overline{D'} \to \overline{D}$ donné par

$$i(p_1, p') = (p_1, i_{p_1}(p'))$$

donne une foliation de $\Sigma_e = i(D_e \times \Sigma')$ dont les feuilles sont des variétés complexes isomorphe à D_e par π_e.

6.13 <u>Lemme</u> Si $f \in L^2(\Sigma_e, d\sigma)$ vérifie les équations de Cauchy-Riemann tangentielle au sens faible, alor pour presque tout $n' \in N(Q')$, $f(i(p_1, n'))$ est holomorphe en p_1.

Comme i n'est pas une application biholomorphe, la structure holomorphe sur D' déduite de celle de la fibre $\pi_e^{-1}(p_1)$ par i_{p_1} varie avec p_1. On notera D'_{p_1} l'espace D' muni de cette structure. (Les différentes structures sont déduites par la conjugaison par B_1 dans $B_e' \simeq D'$. Au point $p_0 = (is_1, 0)$, l'application i_{p_1} est une application holomorphe de D' sur $\pi_e^{-1}(is_1, 0)$ et est donnée par

$$i_0(z_2, u_2 + z') = (is_1 + z_2 - \tfrac{i}{4}[Jz', z'] + z', u_2)\ ).$$

Si F est une fonction holomorphe sur D, on note

$$||F||^2_{H_2, p_1} = \sup_{\tau' \in \Omega'} \int |F(i(p_1, \sigma' + i\tau', 0))|^2 d\sigma'.$$

C'est-à-dire, c'est la norme de Hardy de la fonction F restreinte à la fibre $\pi_e^{-1}(p_1)$, transportée sur D'_{p_1}. On note aussi $H^2_{p_1}$ le sous espace de $L^2(N'(Q_e'))$ obtenu par valeurs au bord des fonctions F appartenant à l'espace de Hardy $H^2_{p_1}$, (il varie avec p_1).

6.14 Notons enfin

$$\mathcal{H}_e(D_e) = \{F \in \mathcal{O}(D_e);\ \int |F(x_1 + iy_1, u_1)|^2 \mu_e(y_1 - Q(u_1, u_1))\, dx_1\, dy_1\, du_1 < +\infty\}.$$

6.15 Théorème

Si $F \in H_1^e$, on a

$$||F||^2 = \int_{D_1} ||F||^2_{H_2,p_1} \, d\mu_1(p_1)$$

et la correspondance $(\tilde{v}_e F)(n,p_1) = (v_e F)(n' \cdot (z_1,u_1))$

$$= (v_e F)(i(p_1,n'))$$

définit une isométrie de H_1^e sur un sous espace propre de $L^2(N_e(Q_e'),\mathcal{H}_e(D_e))$ et pour tout p_1, $(\tilde{v}_e F)(n',p_1)$ appartient à $H^2_{p_1} \subset L^2(N_e(Q_e'))$.

7. Espaces $H_0(R_{\lambda_e})$ et sous espaces invariants de séries principales.

Supposons pour simplifier, λ entier, de sorte que λ définit un caractère de K. Considerons l'espace $\mathcal{O}(\lambda)$ des fonctions ϕ sur $\tilde{G}$ vérifiant

$$\phi(gk) = \lambda(k)^{-1}\phi(g)$$

$$r(X)\cdot\phi = 0, \quad x \in P^-$$

(voir 1.2).

On peut les considérer comme des fonctions sur $GK_{\mathbb{C}}P_- = \exp(D)K_{\mathbb{C}}P_-$ (voir 5.7).

Les fonctions de $L(\lambda)$ s'étendent continûment à $\exp \bar{D}K_{\mathbb{C}}P_-$ (en effet $c_i^{-1}G \subset P_+K_{\mathbb{C}}P_-$).

Si $\phi \in L(\lambda_e)$ et e étant un indice entre 1 et r on peut donc définir

$$(A_e\phi)(g) = \phi(gc_e) \qquad (c_e \text{ défini en } 6.1).$$

7.1 Posons $W_e = c_e(k^C \oplus P^-)$.

On note $c_e(\lambda)$ le caractère de cette sous algèbre de $G^{\mathbb{C}}$ défini par $\langle c_e(\lambda_e),c_e(x)\rangle = \langle\lambda_e,x\rangle$ (λ est étendu trivialement sur P^-). Il est clair que si $\phi \in L(\lambda_e)$, alors $A_e\phi$ vérifie les équations

7.2 $r(X)\cdot(A_e\phi) = -\langle c_e(\lambda_e),x\rangle\phi$, $x \in W_e$. D'autre part soit M le sous groupe de $g \in K$ commutant à tout élément de A, alors c_e

commute aux éléments de M. On a donc aussi

7.3 $(A_e\phi)(gm) = \lambda(m)^{-1}(A_e\phi)(g)$.

A_e étant un opérateur de multiplication à droite (par un élément de $G_{\mathbb{C}}$), commute à l'action à gauche de G. Donc "formellement", A_e est un opérateur d'entrelacement de la représentation "induite holomorphe" par le caractère λ de $k^{\mathbb{C}} + P^-$ et la représentation "C.R. induite" par le caractère $c_e(\lambda)$ de W_e (remarquons que $W_e + \overline{W}_e$ n'est pas une sous algèbre de $G^{\mathbb{C}}$). Via l'isomorphisme P_{λ_e} (5.10), $L(\lambda_e)$ s'identifie à un sous espace de fonctions de $H_0(R_{\lambda_e}) = H_1^e$. Tandis que via un isomorphisme $\tilde{P}_{\lambda_e}$ qu'on décrira les fonctions C^∞ vérifiant les équations (7.2) et (7.3) s'identifie à un sous espace de fonctions sur Σ_e vérifiant les équations de Cauchy-Riemann tangentielles. On montrera que transporté par ces isomorphismes A_e coincide avec l'isométrie $\tilde{V}_e$ du Théorème 6.15, qui est le passage à la valeur au bord; on obtiendra donc un sous espace invariant propre de $L^2(N_e(Q'_e), H_e(\mu_e))$ et on identifiera ce dernier espace à l'espace d'une représentation τ_e d'une série principale associée à un parabolique maximal P_e de $\tilde{G}$.

7.4 Soit $Js'_e = \frac{1}{2}\sum_e^r X_{\gamma_i}$ et soit

$$G = G_e(-1) \oplus G_e(-\tfrac{1}{2}) \oplus G_e(0) \oplus G_e(\tfrac{1}{2}) \oplus G_e(1)$$

la décomposition de G en sous espaces propres pour l'action de Ad Js'_e. On note $P_e = G_e(-1) \oplus G_e(-\frac{1}{2}) \oplus G_e(0)$. Remarquons que

$$G_e(1) = H_1(C'_e)$$

$$G_e(\tfrac{1}{2}) = H'_{1/2}$$

On note $H'^{-}_{1/2} = b^- \cap (H'_{1/2})^{\mathbb{C}}$. On définit $b_e^- = b_e^{\mathbb{C}} \cap b^-$ (6.12). On note V_e le sous groupe $\exp\left(G_e(-\frac{1}{2}) \oplus G_e(-1)\right)$ et $N'_e(Q'_e)$ le sous groupe $\exp\left(G_e(+\frac{1}{2}) \oplus G_e(1)\right)$ dejà introduit et $G_e(0) = \{g \in G; g\cdot Js'_e = Js'_e\}$. $G_e(0)$ est engendré par sa composante connexe et M [1] le groupe $P_e = V_e \cdot G_e(0)$ est un sous groupe parabolique maximal de $\tilde{G}$ d'algèbre de Lie P_e. On a $G = N'\cdot P_e$ à ensemble de mesure nulle près.

On note $P_e^- = c_e(k^{\mathbb{C}} \oplus P^-) \cap P_e^{\mathbb{C}}$

$h_e^- = c_e(k^{\mathbb{C}} \oplus P^-) \cap b^{\mathbb{C}}$

7.5 Lemme .

* On a $W_e = P_e^- \oplus H'^-_{1/2}$ Et une fonction Φ , C^∞ sur $O_e = G\cdot\zeta(c_e) \subset P^+$ vérifie les équations de Cauchy-Riemann tangentielles si et seulement si cette fonction considérée comme fonction sur G est annulée par les champs de vecteurs $r(X)$, pour $X \in W_e$ et est invariante à droite par M

* On a $h_e^- = H_0(\mathbb{C}_e')^{\mathbb{C}} \oplus H'^-_{1/2} \oplus b_e^-$. Une fonction Φ , C^∞ sur $B\cdot(is_e,0) = \Sigma_e \subset H_1^{\mathbb{C}} \oplus H_{1/2}^+$ vérifie les équations de Cauchy-Riemann tangentielles si et seulement si cette fonction considérée comme une fonction sur B est annulée par tous les champs de vecteurs $r(X)$, $X \in h_e^-$.

7.6 Soit $C(\lambda_e;P_e^-;P_e) = \{\phi$ fonction C^∞ sur P_e vérifiant $r(X)\cdot\phi = -\langle\lambda, c_e^{-1}(X)\rangle\phi$, $X \in P_e^-$ et $\phi(gm) = \lambda(m)^{-1}\phi(g)\}$. On voit facilement qu'une telle fonction est entièrement determinée par sa restriction à B_e.

On note $||\phi||^2 = \int_{B_e} |\phi(b_1)|^2 db_1$ et $H(\lambda_e;P_e^-;P_e)$ l'espace de Hilbert des fonctions $\phi \in C(\lambda_e;P_e^-;P_e)$ de norme finie. L'application $(P_{\lambda_e}^1 F)(b_1) = \Phi_{\lambda_e}(b_1)^{-1}\cdot F(\alpha_e(b_1))$ réalise un isomorphisme entre $H_e(\mu_{\lambda_e})$ (voir 6.14) et $H(\lambda_e;P_e^-;P_e)$.

On considère la représentation $\mathcal{T}_e$ de P_e dans $H(\lambda_e;P_e^-;P_e)$ par translations à gauche. Si on note $\rho_e(X) = \frac{1}{2}\operatorname{Tr}_{P_e}$ ad X et aussi ρ_e le caractère correspondant de P_e, alors $\rho_e \otimes \mathcal{T}_e$ est une représentation unitaire irréductible de P_e dans $H(\lambda_e;P_e^-;P_e)$.

7.7 La représentation τ_e en question est la représentation unitaire

$$\operatorname{Ind}\uparrow_{P_e}^{\tilde{G}} \rho_e \otimes \mathcal{T}_e$$

D'après la transitivité des représentations induites holomorphes (ch.5 [2]), τ_e se réalise par translations à gauche dans l'espace $H(\lambda_e;P_e^-;\tilde{G})$, complété des fonctions ϕ:C^∞ sur $\tilde{G}$ vérifiant

$r(X)\cdot\phi = -\langle\lambda,c_e^{-1}(X)\rangle\phi,\quad X \in P_e^-$

$\phi(gm) = \lambda(m)^{-1}\phi(g),\quad m \in M$

pour la norme $||\phi||^2 = \int_{N'\times B_e} |\phi(n'b_1)|^2 dn'db_1$.

7.8 On définit $H(\lambda_e;W_e;G)$ le sous espace complété des fonctions ϕ, C^∞ sur $\tilde{G}$ vérifiant

$$r(X)\cdot\phi = -\langle\lambda,c_e^{-1}(X)\rangle\cdot\phi,\qquad X \in W_e$$
$$\phi(gm) = \lambda(m)^{-1}\phi(g)$$

dans $H(\lambda_e;P_e^-;\tilde{G})$

7.9 Et on définit un isomorphisme $\tilde{P}_{\lambda_e}:L^2(N';H_e(\mu_{\lambda_e})) \to H(\lambda_e;P_e^-;\tilde{G})$ par $(\tilde{P}_{\lambda_e}\phi)(n'b_1) = \phi_{\lambda_e}(b_1)^{-1}\cdot(\phi(n')(\alpha_e(b_1)))$.

Théorème. L'application A_e envoie unitairement $L(\lambda_e)$ dans le sous espace $H(\lambda_e;W_e;\tilde{G})$ de $H(\lambda_e;P_e^-;\tilde{G})$ et se prolonge donc à $H^0(\lambda_e)$.

Le diagrame

$$\begin{array}{ccc} H_0(\lambda_e) & \xrightarrow{A_e} & H(\lambda_e;P_e^-;\tilde{G}) \\ \updownarrow P_{\lambda_e} & & \uparrow \tilde{P}_{\lambda_e} \\ H^0(R_{\lambda_e}) & \xrightarrow{\tilde{v}_e} & L^2(N';H_e(\mu_{\lambda_e})) \end{array}$$

est commutatif, de sorte que $\tilde{v}_e$ est un opérateur d'entrelacement entre la représentation irreductible de $\tilde{G}$ dans $H_0(R_{\lambda_e})$ et une sous représentation de la représentation τ_e.

Si $e \neq 1$, ou si $e = 1$ et G/K n'est pas de type I, alors

A_e est un isomorphisme de $H_0(\lambda_e)$ sur $H(\lambda_e; W_e; \tilde{G})$.

REFERENCES

1. BAILY, W., BOREL, A.:
Compactification of arithmetic quotients of bounded symmetric domains. Annals of Math. (2) 84 , 442-528 (1966).

2. BERNAT, P. , et al. :
Représentations des groupes de Lie résolubles. Paris : Dunod 1972.

3. DIXMIER, J.:
Algèbres enveloppantes. Paris : Gauthier-Villars 1974.

4. GINDIKIN, S. G.:
Analysis on homogeneous domains. Russian Math. Surveys 19, 1-89 (1964).

5. GREENFIELD, S.:
Cauchy-Riemann equations in several variables. Am. S.N.S. Pisa (3) 22, 275-314 (1968).

6. GROSS, K., KUNZE, R.:
"Generalized Bessel transforms and Unitary representations", Harmonic analysis on homogeneous spaces, Proceedings of Symposia in Pure Mathematics, vol XXVI, A.M.S, 1973.

7. HARISH-CHANDRA:
Representations of semi-simple Lie Groups,
IV. Amer. J. Math, 77, 743-777 (1955)
V. Amer. J. Math 78, 1-41 (1956)
VI. Amer. J. Math 78, 564-628 (1956).

8. KNAPP, A., OKAMOTO, K.:
Limits of holomorphic Discrete Series. J. of Funct. Anal. 9, 375-409 (1972)

9. KORANYI, A., STEIN, E.:
H^2 spaces of generalized Half Planes. Studia Mathematica 44, 379-388 (1972)

10. KORANYI, A., WOLF, J.A.:
Realization of Hermitian symmetric spaces as generalized half-planes.
Ann. of Math. 81, 265-288 (1965).

11. KUNZE, R.:
On the irreductibility of certain multiplier representations.
Bull. Amer. Soc. 68 , 93-94 (1962).

12. KUNZE, R. :
Positive Definite Operator-Valued Kernels and Unitary Representations.

13. MOORE, C.C.:
Compactification of symmetric spaces II. Amer. J. Math. 86 , 358-378 (1964).

14. ROSSI, H., VERGNE, M. :
Representations of certain solvable Lie groups on Hilbert spaces of holomorphic functions and the application to the holomorphic discrete series of a semi-simple Lie group. J. Funct. Anal. 13(4), 324-389 (1973).

15. WALLACH, N.:
Induced Representations of Lie algebras, II
Proceedings Amer. Math. Soc. 21 , 161-166 (1969).

16. WALLACH, N.:
Analytic continuation of the discrete series (I), à paraitre.

17. WOLF, J.A.:
Fine structure of Hermitian Symmetric spaces.
"Symmetric Spaces". New-York : Marcel Dekker 1972.

18. WOLF, J.A., KORANYI, A.:
Generalized Cayley transformations of bounded symmetric domains.
Amer. J. Math. 87 , 899-939 (1965).

Université PARIS VII
U.E.R de Mathématiques
2 Place Jussieu
75221 PARIS CEDEX 05

On the Unitarizability of Representations with Highest Weights

Nolan R. Wallach

1. Introduction. In a series of papers ([1], [2], [3]), Harish-Chandra made an extensive study of certain representations of the universal covering group, G, of the group of holomorphic automorphisms of a bounded symmetric domain. The class of representations studied by Harish-Chandra includes the holomorphic discrete series. In this article we study the question of when these representations give rise to unitary representations of G. Harish-Chandra gives a sufficient condition which (of course) includes the holomorphic discrete series and the so-called "limits of discrete series" (see Knapp-Okamoto [8]). However, it is well known that Harish-Chandra's condition is not a necessary condition. Indeed, in Sally [7] a complete analysis of the "analytic continuation of the discrete series "analytic continuation of the discrete series" is given for the universal covering group of SL(2,R) and he finds unitary representations considerably past the "limit of discrete series". In Vergne's article in this volume she describes sufficient conditions (prove jointly with Rossi) for a certain class of these representations. We will see that the Rossi-Vergne condition is "essentially" a necessary one.

One of the main reasons for studying the unitarizability of these representations comes from the study of the topology of Hermitian locally symmetric spaces of negative curvature. Such a space, if it is compact, is of the form $\Gamma\backslash G/K$ with Γ a uniform discrete subgroup of G. If $0 \leq p <$ complex dimension of G/K then the 0,p - Betti number of $\Gamma\backslash G/K$ is a sum of multiplicities in $L^2(\Gamma\backslash G)$ of unitarizable representations of the type studied in this article.

2. Representations with highest weights. Let $\mathfrak{g}$ be a simple Lie algebra of non-compact type over R. Let $\mathfrak{g}_C$ be its complexification. Let $\mathfrak{h}_0 \subset \mathfrak{g}$ be a Cartan subalgebra and let $\mathfrak{h}$ be the complexification of $\mathfrak{h}_0$ in $\mathfrak{g}_C$. Let Δ be the root system of $(\mathfrak{g}_C, \mathfrak{h})$ and let Δ^+

be a system of positive roots for Δ.

Let for $\alpha \in \Delta$, $\mathfrak{g}_\alpha = \{X \in \mathfrak{g}_C \mid [h,X] = \alpha(h)X$ for all $h \in \mathfrak{h}\}$.

Set $\mathfrak{n}^+ = \sum_{\alpha \in \Delta^+} \mathfrak{g}_\alpha$, $\mathfrak{n}^- = \sum_{\alpha \in \Delta^+} \mathfrak{g}_{-\alpha}$.

<u>Definition 2.1.</u> A representation, (π, V) of $\mathfrak{g}_C$ is said to have highest weight $\Lambda \in \mathfrak{h}^*$ if there exists $v_o \in V$, $v_o \neq 0$, so that

1) If $h \in \mathfrak{h}$ then $\pi(h)v_o = \Lambda(h)v_o$,

2) $\pi(X)v_o = 0$ if $X \in \mathfrak{n}^+$,

3) $\pi(U(\mathfrak{g}_C))v_o = V$

(here $U(\mathfrak{g}_C)$ denotes the universal enveloping algebra of $\mathfrak{g}_C$).

<u>Definition 2.2.</u> Let (π, V) be a representation of $\mathfrak{g}_C$. Then (π, V) is said to be $\mathfrak{g}$-unitarizable if there exists a positive definite inner product on V, $\langle\ ,\ \rangle$, so that if X is in $\mathfrak{g}$ then

$\langle \pi(X)v, w\rangle = -\langle v, \pi(X)w\rangle$ for all v, w in V.

<u>Lemma 2.3.</u> If (π, V) has highest weight Λ and is $\mathfrak{g}$-unitarizable then (π, V) is irreducible.

If $\Lambda = 0$ then the irreducible representation of $\mathfrak{g}_C$ with highest weight Λ is the trivial representation which is obviously $\mathfrak{g}$-unitarizable. A reasonable question is: For what $\mathfrak{g}$, $\mathfrak{h}_o$, Δ^+ does there exist $\Lambda \neq 0$ so that the irreducible representation with highest weight Λ is $\mathfrak{g}$-unitarizable? The complete answer to this question is due to Harish-Chandra.

<u>Theorem 2.4 (Harish-Chandra [1], [3]).</u> A necessary and sufficient condition that there exists $\Lambda \neq 0$ so that a representation of $\mathfrak{g}_C$ with highest weight Λ is $\mathfrak{g}$-unitarizable is

1) There exists a Cartan decomposition of $\mathfrak{g}$, $\mathfrak{g} = \mathfrak{k} \oplus \mathfrak{p}$ so that $\mathfrak{h}_o \subset \mathfrak{k}$.

2) $[\mathfrak{k},\mathfrak{k}] \neq \mathfrak{k}$ and $\mathfrak{k} = RiH \oplus [\mathfrak{k},\mathfrak{k}]$ with $H \in \mathfrak{h}$. Furthermore if $\alpha \in \Delta^+$ then $\alpha(H) = 0$ or 1.

If $\mathfrak{g}$ satisfies 1), 2) above and if X is the symmetric space corresponding to $\mathfrak{g}$ then X is Hermitian symmetric. The sufficiency of 1), 2) is proved in Harish-Chandra [3] using the existence of the holomorphic discrete series. A completely algebraic proof of this result can be found in Wallach [11].

A natural question is:

2.5. Assuming $(\mathfrak{g}, \mathfrak{h}_o, \Delta^+)$ satisfy 1), 2) of theorem 2.4 for what Λ is the irreducible representation with highest weight Λ $\mathfrak{g}$-unitarizable?

In Vergne's article in this volume a sufficient condition is given. We recall this condition (due to N. Conze).
Let $\Delta_K^+ = \{\alpha \in \Delta^+ \mid \alpha(H) = 0\}$, $\Delta_P^+ = \{\alpha \in \Delta^+ \mid \alpha(H) = 1\}$. ($\Delta_K^+ \cup \Delta_P^+ = \Delta^+$ by 2.4 2).) Let $\rho = \frac{1}{2}\sum_{\alpha \in \Delta^+} \alpha$. Let β be the largest element of Δ.

Theorem 2.6 (N. Conze). The irreducible representation with highest weight Λ is $\mathfrak{g}$-unitarizable if the following two conditions are satisfied:

1) If $\alpha \in \Delta_K^+$ then $2\langle\Lambda,\alpha\rangle/\langle\alpha,\alpha\rangle$ is a non-negative integer,

2) $2\langle\Lambda+\rho,\beta\rangle/\langle\beta,\beta\rangle$ is a real number ≤ 1.

Theorem 2.6 leads to the following question:

2.7 Is 1),2) a necessary condition?

We shall see in the next section that the answer to 2.7 is no. There is evidence that "generically" the answer to 2.7 is yes. We will discuss this in section 4.

3. The case of line bundles. We retain the notation of section 2. We assume that conditions 1) and 2) of Theorem 2.4 are satisfied. Let $\Pi = \{\alpha_1, \ldots, \alpha_n\}$ be the simple roots of Δ^+. We can (and do) assume that $\Pi \cap \Delta_p^+ = \{\alpha_1\}$. Let $\Lambda_1 \in \mathfrak{h}^*$ be defined by $2\langle\Lambda_1, \alpha_j\rangle / \langle\alpha_j, \alpha_j\rangle = \delta_{1,j}$. Then, if $\Lambda \in \mathfrak{h}^*$, $\Lambda = z\Lambda_1 + \Lambda_o$ with $\langle\Lambda_o, \alpha_1\rangle = 0$.

Lemma 3.1 (Harish-Chandra [1]). A necessary condition that the irreducible representation with highest weight $\Lambda = z\Lambda_1 + \Lambda_o$ be $\mathfrak{g}$-unitarizable is

1) Λ_o is dominant integral,

2) $z \in R$, $z < 0$.

We will give a necessary and sufficient condition in the case $\Lambda_o = 0$. We use some notation from Vergne's article in these proceedings. Let $\gamma_1, \ldots, \gamma_r$ be as in section 4.1 of Vergne's article and let p be as in section 4.6 of Vergne's article.

Theorem 3.2. A necessary and sufficient condition for the irreducible representation with highest weight $z\Lambda_1$ to be $\mathfrak{g}$-unitarizable is that $z \in \mathbb{R}$ and

1) $z < -(r-1)p/2$

or

2) $z = -(j-1)p/2$ for $j=1, \ldots, r$.

The sufficiency of $z \leqslant -(r-1)p/2$ is due independently to the present author and to Rossi and Vergne (see Theorem 4.7 of Vergne's article in these proceedings).

Our proof of Theorem 3.2 is algebraic and uses a theorem of W. Schmid [8] that gives the decomposition of the action of K on the symmetric algebra on $\mathfrak{p}^- = \sum_{\alpha \in \Delta_p^+} \mathfrak{g}_\alpha$ in terms of $\gamma_1, \ldots, \gamma_r$. The details can be found in [11].

We note that Theorem 3.2 implies that 2.7 has a negative answer.

In the next section we give an example which indicates that the case $\Lambda = z\Lambda_1$ is an isolated phenomenum.

<u>4. The universal covering group of SU(n,1).</u> Let $\mathfrak{g}$ be the Lie algebra of SU(n,1). Then $\mathfrak{g} \subset \mathfrak{g}_C = sl(n+1,C) = \{X \mid X$ an nxn matrix of trace zero$\}$. We choose $\mathfrak{h}_o$ and Δ^+ satisfying the conditions of 2.4. We order Π so that α_1 is as in section 3. Then Π has Dynkin diagram

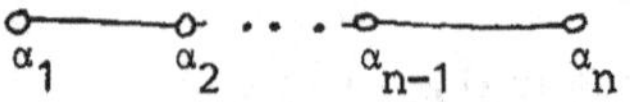

Let $\Lambda_n \in \mathfrak{h}^*$ be defined by $2\langle\Lambda_n, \alpha_j\rangle / \langle\alpha_j, \alpha_j\rangle = \delta_{n,j}$. We look at linear forms of the form $\Lambda = z\Lambda_1 + k\Lambda_n$. Lemma 3.1 says that we should only consider $z \in \mathbb{R}$, $z < 0$ and $k \in \mathbb{Z}$, $k \geq 0$.

<u>Theorem 4.1.</u> A necessary and sufficient condition for $\Lambda = z\Lambda_1 + k\Lambda_n$ to be the highest weight of a $\mathfrak{g}$-unitarizable representation is

1) If $k=0$, $z \leq 0$,
2) If $k > 0$ then $2\langle\Lambda + \rho, \beta\rangle / \langle\beta, \beta\rangle \leq 1$.

Note that if n=1 or 2 then Theorem 4.4 completely answers question 2.5. Our proof of Theorem 4.1 is an extension of the techniques in Johnson-Wallach [5] (a diligent reader of that paper can find the result for the case $\Lambda = k(\Lambda_1 + \Lambda_n)$).

REFERENCES

[1] Harish-Chandra, Representations of semi-simple Lie groups IV, Amer. J. Math.,77(1955),743-777.

[2] , Representations of semi-simple Lie groups V, Amer. J. Math.,78(1956),1-41.

[3] , Representations of semi-simple Lie groups VI, Amer. J. Math.,78(1956),564-628.

[4] R. Hotta and N.R. Wallach, On Matsushima's formula for the Betti numbers of a locally symmetric space, to appear.

[5] K. Johnson and N.R. Wallach, Composition series and intertwining operators for the spherical principal series I, to appear.

[6] A. Knapp and K. Okamoto, Limits of holomorphic discrete series, Jour. of Func. Anal.,9(1972),375-409.

[7] P. Sally, Analytic continuation of the irreducible unitary representations of the universal covering group of SL(2,R), Memoirs of the Amer. Math. Soc.,No.69, 1967.

[8] W. Schmid, Die randwerte holomorpher Funktionen auf hermitesch symmetrischen Räumen, Inventiones Math.,9(1969),61-80.

[9] M. Vergne, Article in these proceedings.

[10] N.R. Wallach, The analytic continuation of the discrete series I, to appear.

[11] N.R. Wallach, The analytic continuation of the discrete series II, to appear.

Department of Mathematics
Rutgers University
New-Brunswick, N.J. 08903 / USA

Vol. 309: D. H. Sattinger, Topics in Stability and Bifurcation Theory. VI, 190 pages. 1973. DM 20,-

Vol. 310: B. Iversen, Generic Local Structure of the Morphisms in Commutative Algebra. IV, 108 pages. 1973. DM 18,-

Vol. 311: Conference on Commutative Algebra. Edited by J. W. Brewer and E. A. Rutter. VII, 251 pages. 1973. DM 24,-

Vol. 312: Symposium on Ordinary Differential Equations. Edited by W. A. Harris, Jr. and Y. Sibuya. VIII, 204 pages. 1973. DM 22,-

Vol. 313: K. Jörgens and J. Weidmann, Spectral Properties of Hamiltonian Operators. III, 140 pages. 1973. DM 18,-

Vol. 314: M. Deuring, Lectures on the Theory of Algebraic Functions of One Variable. VI, 151 pages. 1973. DM 18,-

Vol. 315: K. Bichteler, Integration Theory (with Special Attention to Vector Measures). VI, 357 pages. 1973. DM 29,-

Vol. 316: Symposium on Non-Well-Posed Problems and Logarithmic Convexity. Edited by R. J. Knops. V, 176 pages. 1973. DM 20,-

Vol. 317: Séminaire Bourbaki – vol. 1971/72. Exposés 400–417. IV, 361 pages. 1973. DM 29,-

Vol. 318: Recent Advances in Topological Dynamics. Edited by A. Beck. VIII, 285 pages. 1973. DM 27,-

Vol. 319: Conference on Group Theory. Edited by R. W. Gatterdam and K. W. Weston. V, 188 pages. 1973. DM 20,-

Vol. 320: Modular Functions of One Variable I. Edited by W. Kuyk. V, 195 pages. 1973. DM 20,-

Vol. 321: Séminaire de Probabilités VII. Edité par P. A. Meyer. VI, 322 pages. 1973. DM 29,-

Vol. 322: Nonlinear Problems in the Physical Sciences and Biology. Edited by I. Stakgold, D. D. Joseph and D. H. Sattinger. VIII, 357 pages. 1973. DM 29,-

Vol. 323: J. L. Lions, Perturbations Singulières dans les Problèmes aux Limites et en Contrôle Optimal. XII, 645 pages. 1973. DM 46,-

Vol. 324: K. Kreith, Oscillation Theory. VI, 109 pages. 1973. DM 18,-

Vol. 325: C.-C. Chou, La Transformation de Fourier Complexe et L'Equation de Convolution. IX, 137 pages. 1973. DM 18,-

Vol. 326: A. Robert, Elliptic Curves. VIII, 264 pages. 1973. DM 24,-

Vol. 327: E. Matlis, One-Dimensional Cohen-Macaulay Rings. XII, 157 pages. 1973. DM 20,-

Vol. 328: J. R. Büchi and D. Siefkes, The Monadic Second Order Theory of All Countable Ordinals. VI, 217 pages. 1973. DM 22,-

Vol. 329: W. Trebels, Multipliers for (C, α)-Bounded Fourier Expansions in Banach Spaces and Approximation Theory. VII, 103 pages. 1973. DM 18,-

Vol. 330: Proceedings of the Second Japan-USSR Symposium on Probability Theory. Edited by G. Maruyama and Yu. V. Prokhorov. VI, 550 pages. 1973. DM 40,-

Vol. 331: Summer School on Topological Vector Spaces. Edited by L. Waelbroeck. VI, 226 pages. 1973. DM 22,-

Vol. 332: Séminaire Pierre Lelong (Analyse) Année 1971-1972. V, 131 pages. 1973. DM 18,-

Vol. 333: Numerische, insbesondere approximationstheoretische Behandlung von Funktionalgleichungen. Herausgegeben von R. Ansorge und W. Törnig. VI, 296 Seiten. 1973. DM 27,-

Vol. 334: F. Schweiger, The Metrical Theory of Jacobi-Perron Algorithm. V, 111 pages. 1973. DM 18,-

Vol. 335: H. Huck, R. Roitzsch, U. Simon, W. Vortisch, R. Walden, B. Wegner und W. Wendland, Beweismethoden der Differentialgeometrie im Großen. IX, 159 Seiten. 1973. DM 20,-

Vol. 336: L'Analyse Harmonique dans le Domaine Complexe. Edité par E. J. Akutowicz. VIII, 169 pages. 1973. DM 20,-

Vol. 337: Cambridge Summer School in Mathematical Logic. Edited by A. R. D. Mathias and H. Rogers. IX, 660 pages. 1973. DM 46,-

Vol. 338: J. Lindenstrauss and L. Tzafriri, Classical Banach Spaces. IX, 243 pages. 1973. DM 24,-

Vol. 339: G. Kempf, F. Knudsen, D. Mumford and B. Saint-Donat, Toroidal Embeddings I. VIII, 209 pages. 1973. DM 22,-

Vol. 340: Groupes de Monodromie en Géométrie Algébrique. (SGA 7 II). Par P. Deligne et N. Katz. X, 438 pages. 1973. DM 44,-

Vol. 341: Algebraic K-Theory I, Higher K-Theories. Edited by H. Bass. XV, 335 pages. 1973. DM 29,-

Vol. 342: Algebraic K-Theory II, "Classical" Algebraic K-Theory, and Connections with Arithmetic. Edited by H. Bass. XV, 527 pages. 1973. DM 40,-

Vol. 343: Algebraic K-Theory III, Hermitian K-Theory and Geometric Applications. Edited by H. Bass. XV, 572 pages. 1973. DM 40,-

Vol. 344: A. S. Troelstra (Editor), Metamathematical Investigation of Intuitionistic Arithmetic and Analysis. XVII, 485 pages. 1973. DM 38,-

Vol. 345: Proceedings of a Conference on Operator Theory. Edited by P. A. Fillmore. VI, 228 pages. 1973. DM 22,-

Vol. 346: Fučík et al., Spectral Analysis of Nonlinear Operators. II, 287 pages. 1973. DM 26,-

Vol. 347: J. M. Boardman and R. M. Vogt, Homotopy Invariant Algebraic Structures on Topological Spaces. X, 257 pages. 1973. DM 24,-

Vol. 348: A. M. Mathai and R. K. Saxena, Generalized Hypergeometric Functions with Applications in Statistics and Physical Sciences. VII, 314 pages. 1973. DM 26,-

Vol. 349: Modular Functions of One Variable II. Edited by W. Kuyk and P. Deligne. V, 598 pages. 1973. DM 38,-

Vol. 350: Modular Functions of One Variable III. Edited by W. Kuyk and J.-P. Serre. V, 350 pages. 1973. DM 26,-

Vol. 351: H. Tachikawa, Quasi-Frobenius Rings and Generalizations. XI, 172 pages. 1973. DM 20,-

Vol. 352: J. D. Fay, Theta Functions on Riemann Surfaces. V, 137 pages. 1973. DM 18,-

Vol. 353: Proceedings of the Conference on Orders, Group Rings and Related Topics. Organized by J. S. Hsia, M. L. Madan and T. G. Ralley. X, 224 pages. 1973. DM 22,-

Vol. 354: K. J. Devlin, Aspects of Constructibility. XII, 240 pages. 1973. DM 24,-

Vol. 355: M. Sion, A Theory of Semigroup Valued Measures. V, 140 pages. 1973. DM 18,-

Vol. 356: W. L. J. van der Kallen, Infinitesimally Central-Extensions of Chevalley Groups. VII, 147 pages. 1973. DM 18,-

Vol. 357: W. Borho, P. Gabriel und R. Rentschler, Primideale in Einhüllenden auflösbarer Lie-Algebren. V, 182 Seiten. 1973. DM 20,-

Vol. 358: F. L. Williams, Tensor Products of Principal Series Representations. VI, 132 pages. 1973. DM 18,-

Vol. 359: U. Stammbach, Homology in Group Theory. VIII, 183 pages. 1973. DM 20,-

Vol. 360: W. J. Padgett and R. L. Taylor, Laws of Large Numbers for Normed Linear Spaces and Certain Fréchet Spaces. VI, 111 pages. 1973. DM 18,-

Vol. 361: J. W. Schutz, Foundations of Special Relativity: Kinematic Axioms for Minkowski Space Time. XX, 314 pages. 1973. DM 26,-

Vol. 362: Proceedings of the Conference on Numerical Solution of Ordinary Differential Equations. Edited by D. Bettis. VIII, 490 pages. 1974. DM 34,-

Vol. 363: Conference on the Numerical Solution of Differential Equations. Edited by G. A. Watson. IX, 221 pages. 1974. DM 20,-

Vol. 364: Proceedings on Infinite Dimensional Holomorphy. Edited by T. L. Hayden and T. J. Suffridge. VII, 212 pages. 1974. DM 20,-

Vol. 365: R. P. Gilbert, Constructive Methods for Elliptic Equations. VII, 397 pages. 1974. DM 26,-

Vol. 366: R. Steinberg, Conjugacy Classes in Algebraic Groups (Notes by V. V. Deodhar). VI, 159 pages. 1974. DM 18,-

Vol. 367: K. Langmann und W. Lütkebohmert, Cousinverteilungen und Fortsetzungssätze. VI, 151 Seiten. 1974. DM 16,-

Vol. 368: R. J. Milgram, Unstable Homotopy from the Stable Point of View. V, 109 pages. 1974. DM 16,-

Vol. 369: Victoria Symposium on Nonstandard Analysis. Edited by A. Hurd and P. Loeb. XVIII, 339 pages. 1974. DM 26,-

Vol. 370: B. Mazur and W. Messing, Universal Extensions and One Dimensional Crystalline Cohomology. VII, 134 pages. 1974. DM 16,–

Vol. 371: V. Poenaru, Analyse Différentielle. V, 228 pages. 1974. DM 20,–

Vol. 372: Proceedings of the Second International Conference on the Theory of Groups 1973. Edited by M. F. Newman. VII, 740 pages. 1974. DM 48,–

Vol. 373: A. E. R. Woodcock and T. Poston, A Geometrical Study of the Elementary Catastrophes. V, 257 pages 1974 DM 22,–

Vol. 374: S. Yamamuro, Differential Calculus in Topological Linear Spaces IV, 179 pages. 1974 DM 18,–

Vol. 375: Topology Conference 1973. Edited by R. F. Dickman Jr. and P. Fletcher. X, 283 pages. 1974. DM 24,–

Vol 376: D B. Osteyee and I J Good, Information, Weight of Evidence, the Singularity between Probability Measures and Signal Detection. XI, 156 pages. 1974. DM 16.–

Vol. 377: A. M. Fink, Almost Periodic Differential Equations. VIII, 336 pages. 1974 DM 26,–

Vol. 378: TOPO 72 – General Topology and its Applications. Proceedings 1972. Edited by R. Alò, R. W. Heath and J. Nagata. XIV, 651 pages. 1974. DM 50,–

Vol. 379: A. Badrikian et S. Chevet, Mesures Cylindriques, Espaces de Wiener et Fonctions Aléatoires Gaussiennes. X, 383 pages. 1974. DM 32,–

Vol. 380: M. Petrich, Rings and Semigroups. VIII, 182 pages. 1974. DM 18,–

Vol. 381: Séminaire de Probabilités VIII. Edité par P A. Meyer IX, 354 pages. 1974. DM 32,–

Vol. 382: J. H. van Lint, Combinatorial Theory Seminar Eindhoven University of Technology VI, 131 pages. 1974. DM 18,–

Vol. 383: Séminaire Bourbaki – vol. 1972/73. Exposés 418-435 IV, 334 pages. 1974. DM 30,–

Vol. 384: Functional Analysis and Applications, Proceedings 1972. Edited by L. Nachbin. V, 270 pages. 1974. DM 22,–

Vol. 385: J. Douglas Jr. and T. Dupont, Collocation Methods for Parabolic Equations in a Single Space Variable (Based on C^1-Piecewise-Polynomial Spaces). V, 147 pages. 1974. DM 16,–

Vol. 386: J. Tits, Buildings of Spherical Type and Finite BN-Pairs. IX, 299 pages. 1974. DM 24,–

Vol. 387: C. P. Bruter, Eléments de la Théorie des Matroïdes. V, 138 pages. 1974. DM 18,–

Vol. 388: R. L. Lipsman, Group Representations. X, 166 pages. 1974. DM 20,–

Vol. 389: M.-A. Knus et M. Ojanguren, Théorie de la Descente et Algèbres d' Azumaya. IV, 163 pages. 1974. DM 20,–

Vol. 390: P. A. Meyer, P. Priouret et F. Spitzer, Ecole d'Eté de Probabilités de Saint-Flour III – 1973. Edité par A. Badrikian et P.-L. Hennequin. VIII, 189 pages. 1974. DM 20,–

Vol. 391: J. Gray, Formal Category Theory: Adjointness for 2-Categories. XII, 282 pages. 1974. DM 24,–

Vol. 392: Géométrie Différentielle, Colloque, Santiago de Compostela, Espagne 1972. Edité par E. Vidal. VI, 225 pages. 1974. DM 20,–

Vol. 393: G. Wassermann, Stability of Unfoldings. IX, 164 pages. 1974. DM 20,–

Vol. 394: W. M. Patterson 3rd, Iterative Methods for the Solution of a Linear Operator Equation in Hilbert Space – A Survey. III, 183 pages. 1974. DM 20,–

Vol. 395: Numerische Behandlung nichtlinearer Integrodifferential- und Differentialgleichungen. Tagung 1973. Herausgegeben von R. Ansorge und W. Törnig. VII, 313 Seiten 1974. DM 28,–

Vol. 396: K. H. Hofmann, M. Mislove and A. Stralka, The Pontryagin Duality of Compact O-Dimensional Semilattices and its Applications. XVI, 122 pages. 1974. DM 18,–

Vol 397: T. Yamada, The Schur Subgroup of the Brauer Group V, 159 pages. 1974. DM 18,–

Vol. 398: Théories de l'Information, Actes des Rencontres de Marseille-Luminy, 1973. Edité par J. Kampé de Fériet et C. Picard. XII, 201 pages. 1974. DM 23,–

Vol. 399: Functional Analysis and its Applications, Proceedings 1973. Edited by H. G. Garnir, K. R. Unni and J. H. Williamson. XVII, 569 pages. 1974. DM 44,–

Vol. 400: A Crash Course on Kleinian Groups – San Francisco 1974. Edited by L. Bers and I. Kra. VII, 130 pages. 1974. DM 18,–

Vol. 401: F. Atiyah, Elliptic Operators and Compact Groups. V, 93 pages. 1974. DM 18,–

Vol. 402: M. Waldschmidt, Nombres Transcendants. VIII, 277 pages. 1974. DM 25,–

Vol. 403: Combinatorial Mathematics – Proceedings 1972. Edited by D. A. Holton. VIII, 148 pages. 1974. DM 18,–

Vol. 404: Théorie du Potentiel et Analyse Harmonique. Edité par J. Faraut. V, 245 pages. 1974. DM 25,–

Vol. 405: K. Devlin and H. Johnsbråten, The Souslin Problem. VIII, 132 pages. 1974. DM 18,–

Vol. 406: Graphs and Combinatorics – Proceedings 1973. Edited by R. A. Bari and F. Harary. VIII, 355 pages. 1974. DM 30,–

Vol. 407: P. Berthelot, Cohomologie Cristalline des Schémas de Caracteristique p > o. VIII, 598 pages. 1974. DM 44,–

Vol. 408: J. Wermer, Potential Theory. VIII, 146 pages. 1974. DM 18,–

Vol. 409: Fonctions de Plusieurs Variables Complexes, Séminaire François Norguet 1970–1973. XIII, 612 pages. 1974. DM 47,–

Vol. 410: Séminaire Pierre Lelong (Analyse) Année 1972–1973. VI, 181 pages. 1974. DM 18,–

Vol. 411: Hypergraph Seminar. Ohio State University, 1972. Edited by C. Berge and D. Ray-Chaudhuri. IX, 287 pages. 1974. DM 28,–

Vol. 412: Classification of Algebraic Varieties and Compact Complex Manifolds. Proceedings 1974. Edited by H. Popp. V, 333 pages. 1974. DM 30,–

Vol. 413: M. Bruneau, Variation Totale d'une Fonction. XIV, 332 pages. 1974. DM 30,–

Vol. 414: T. Kambayashi, M. Miyanishi and M. Takeuchi, Unipotent Algebraic Groups VI, 165 pages 1974 DM 20,–

Vol. 415: Ordinary and Partial Differential Equations, Proceedings of the Conference held at Dundee, 1974. XVII, 447 pages. 1974. DM 37,–

Vol. 416: M. E. Taylor, Pseudo Differential Operators. IV, 155 pages. 1974. DM 18,–

Vol. 417: H. H. Keller, Differential Calculus in Locally Convex Spaces. XVI, 131 pages. 1974. DM 18,–

Vol. 418: Localization in Group Theory and Homotopy Theory and Related Topics Battelle Seattle 1974 Seminar. Edited by P. J. Hilton. VI, 171 pages. 1974. DM 20,–

Vol. 419: Topics in Analysis – Proceedings 1970. Edited by O. E. Lehto, I. S Louhivaara, and R. H. Nevanlinna. XIII, 391 pages. 1974. DM 35,–

Vol. 420: Category Seminar. Proceedings, Sydney Category Theory Seminar 1972/73. Edited by G. M. Kelly. VI, 375 pages. 1974. DM 32,–

Vol. 421: V. Poénaru, Groupes Discrets VI, 216 pages. 1974. DM 23,–

Vol. 422: J.-M. Lemaire, Algèbres Connexes et Homologie des Espaces de Lacets. XIV, 133 pages 1974 DM 23,–

Vol. 423: S. S. Abhyankar and A. M. Sathaye, Geometric Theory of Algebraic Space Curves. XIV, 302 pages. 1974. DM 28,–

Vol. 424: L. Weiss and J. Wolfowitz, Maximum Probability Estimators and Related Topics. V, 106 pages. 1974. DM 18,–

Vol. 425: P. R. Chernoff and J. E. Marsden, Properties of Infinite Dimensional Hamiltonian Systems. IV, 160 pages 1974. DM 20,–

Vol. 426: M. L Silverstein, Symmetric Markov Processes. IX, 287 pages 1974. DM 28,–

Vol. 427: H Omori, Infinite Dimensional Lie Transformation Groups. XII, 149 pages. 1974 DM 18,–

Vol. 428: Algebraic and Geometrical Methods in Topology, Proceedings 1973. Edited by L. F. McAuley. XI, 280 pages. 1974. DM 28,–